Wildland Fuel Fundamentals and Application

Robert E. Keane

Wildland Fuel Fundamentals and Applications

Springer

Robert E. Keane
Missoula Fire Sciences Laboratory
USDA Forest Service
Rocky Mountain Research Station
Missoula, Montana
USA

ISBN 978-3-319-09014-6 ISBN 978-3-319-09015-3 (eBook)
DOI 10.1007/978-3-319-09015-3
Springer Cham Heidelberg New York Dordrecht London

Library of Congress Control Number: 2014952656

Printed on acid-free paper

Springer is part of Springer Science+Business Media (www.springer.com)

I dedicate this book to my late parents Bob and Betty Keane

Preface

Wildland fire management throughout most of the civilized world is at a critical juncture on how it spends the scant money allocated toward resource management.

These funds could either go toward expensive and largely ineffective wildfire suppression or toward a more holistic and long-term approach that manages wildland fire primarily as an ecological resource with special consideration when fire impacts people and property.

Fire suppression costs are spiraling out of control, often exceeding the value of the land that burned where suppression was attempted. An unfortunate consequence of fighting fires is that even more biomass accumulates, creating conditions that foster even more fires of increased severity. More firefighters will be asked to fight these larger and more dangerous fires, putting themselves at an even greater risk for injury and fatality. The solution to this fire management conundrum will be extremely complex and require consideration of interacting social, political, and ecological issues across multiple time, space, and organizational scales. At the heart of this wildland fire dilemma are wildland fuels.

Wildland fuels may be the most important consideration in fire management, not just because they are important inputs for predicting fire behavior (i.e., how fast and intense a fire gets), but also because fuels are the only factor that can effectively be controlled by direct and indirect management manipulation. Each year, tens if not hundreds of millions of dollars are spent by federal, private, and public organizations on treatments meant to manipulate fuels so that fire spread is reduced and fire intensity is lessened to save property and lives. The amounts and kinds of fuels burned in a fire dictate how dangerous a fire can get, how much smoke is generated, and how ecosystems and society responds. By modifying the fuelbed, managers hope to minimize adverse consequences to protect ecosystems, people, and property. A comprehensive description of wildland fuels is vital for crafting effective treatments and designing novel ways to manage wildfires.

Perhaps an even better reason for the importance of wildland fuels in fire management fire management is that they are the major link between fire behavior and fire ecology. To understand fuel dynamics, we must first understand the ecology of fuels and what happens to the ecosystem before, during, and after fuels burn. Fire behavior is often viewed as the main driver of fire management. This is because most of the funding for fire management is spent trying to extinguish fires rather than managing fuels before a fire occurs. Effective fuel management demands a consideration of both fire behavior and ecological effects. Therefore, all enlightened solutions to most fire management problems will be found in fire ecology rather than fire behavior, and that the principle driver in most fire effects are the combustion of fuels. Long-term solutions to our fire management problems will only be achieved through a comprehensive understanding of the ecology of wildland so that innovative strategies can be crafted to balance ecological, economic, societal and political concerns.

It is somewhat puzzling why more research hasn't been done to fully understand wildland fuel ecology considering the great importance of wildland fuels to fire management. Wildland fuelbeds are amazingly complex, composed of many types of fuels with each fuel type characterized by diverse physical and chemical properties. Yet, most fire and fuel studies and management applications try to simplify fuels rather than attempt to understand their complexity. This over-simplification of information often introduces additional uncertainties that may compromise the assessment of fuels in various research and management applications. Moreover, most information on fuels is scattered across disparate journals, government publications, and websites making it difficult for fire managers to fully understand fuel dynamics without expending a great deal of their precious time. What is needed is a synthesis of fuels information in the context of ecology that can be used to understand basic fuels characteristics to objectively plan, implement, and evaluate results of fire research and management applications.

This book is an attempt to consolidate general introductory material about wildland fuels into a cohesive synthesis that can be used to understand and manage them. It is also meant as a guide for understanding the particular characteristics of fuels so that when fuels data are entered into fire management computer applications, the user can interpret the results in an appropriate context. The intended audience includes students and novice fire professionals who want to understand how fuel is used in fire management applications to better interpret fire simulation results and veteran fire practitioners who wish to have a better context in which to understand their fire analyses. I hope that readers will find that this book provides critical information about fundamental fuel properties and their applications so that fuels can be objectively described, sampled, classified, and mapped using the most appropriate techniques.

This book is divided into two sections: "Fuel Fundamentals" and "Fuel Applications." "Fuel Fundamentals" contains six chapters describing the basic concepts that are used to describe, quantify, and apply fuel information in fire management. The first chapter introduces wildland fuels by detailing their terminology, history and background. Chapter 2 describes how fuels are used in fire modeling applications and details their properties. Chapters 3 and 4 describe surface and canopy fuels, respectively. Chapter 5 presents information on fuel moisture, its measurement, and its interpretation. Chapter 6 ties introductory material together by presenting material on the dynamics of surface and canopy fuels, fuel ecology, and disturbance effects. The chapter also discusses the importance of knowledge of landscape ecology and the concept of scale in understanding fuel dynamics. The term "fuel dynamics" refers to how fuels change over time and space under the processes of deposition and decomposition. These processes are important in evaluating the longevity and effectiveness of fuel treatments.

"Fuel Applications" is a synthesis of currently available applications that use fuel information for fire and fuel management activities. Chapter 7 introduces the reader to the complex world of fuel classifications by summarizing and comparing fuel description systems used today. Chapter 8 tackles fuel sampling and describes the methods and techniques that are commonly used to quantify surface and canopy fuels in the field. Because spatial fuels data layers are critical for most fire analyses, Chapter 9 introduces the reader to the approaches and challenges in mapping fuels. Chapter 10 discusses important fuel concepts used in fire management such as flammability, hazard, and fuel treatments through the lens of fire ecology rather than behavior.

It is also important to know what this book doesn't cover. Most important, this book does not cover fuel treatments. The planning, design, and implementation of fuel treatments were not included because the vast approaches, types, and intensities of fuel treatments almost always demand a local context. That is the subject of another book or series of books. This book also does not detail how to create the applications discussed in the second section. Appropriate fuel sampling designs, classification procedures, and mapping techniques are found in the publications cited in this book and again must be designed around local conditions. This book also doesn't provide instructions on how to use fuel data in fire management applications; those instructions are contained in the manuals for the applications.

I have attempted to provide fuels information from countries all over the world, but this book is admittedly North American in focus, using concepts, examples, and material mostly from the United States and Canada. In addition, this book focuses mainly on fuel loading (mass per unit area) and less so on other important fuel properties, such as heat content, bulk density, and surface-area-to-volume ratios. This is primarily because loading best relates fire ecology to fire behavior and, more pragmatically, most of the literature in wildland fuel science concerns loading. Fuel moisture is also quite important in fire management, and while most of the principles presented in this book are in the context of loading, they could easily be expanded to describe fuel moisture. Terminology is quite important in the fire management community; consistency and consensus in terminology minimizes confusion, facilitates communication, and provides for appropriate descriptions. In this book, terminology is based on the 2013 National Wildfire Coordinating Group (NWCG) glossary (http://www.nwcg.gov/pms/pubs/glossary/).

I wish to acknowledge the many people who extensively reviewed chapters of this book. I thank Colin Hardy, Faith Ann Heisch, Eva Karau, Kris Lee, Signe Leirfallom, Rachel Loehman, Duncan Lutes, Pam Sikkink from the Missoula Fire Sciences Laboratory, Rocky Mountain Research Station, US Forest Service; Matt Rollins,

US Geological Survey; Roger Ottmar at the US Forest Service Seattle Fire Sciences Laboratory; Elizabeth Reinhardt, US Forest Service Fire and Aviation Management Washinton Office; David Affleck, Carl Seielstad, and Casey Teske, University of Montana Missoula Montana; Kevin Ryan, US Forest Service (retired); Marty Alexander, Alberta Canada; Valentina Bacciu Euro-Mediterranean Center on Climate Change Sardinia Italy.

Contents

Section I Fuel Fundamentals

1 Introduction 3

2 Fundamentals 15

3 Surface and Ground Fuels 39

4 Canopy Fuels 57

5 Fuel Moisture 71

6 Ecology 83

Section II Fuel Applications

7 Fuel Classifications 111

8 Fuel Sampling 125

9 Fuel Mapping 153

10 Fuel Concepts 175

Glossary, Abbreviations and Acronyms 185

Index 189

Section I
Fuel Fundamentals

Chapter 1
Introduction

> *There is nothing more difficult to take in hand, more perilous to conduct or more uncertain in its success than ...the introduction of a new order of things.*
> Niccolo Machiavelli

1.1 What Are Fuels?

Wildland fuels are named for the role they play when they are burned by a wildland fire. In a combustion science context, fuels are any combustible material (NWCG 2006). In an ecological context, these combustible materials are the live and dead organic matter that ecologists call *biomass*. Therefore, in this book, fuel is biomass. Some may feel that there are biomass pools that rarely burn, such as large tree boles and snags, and there are some biomass pools that are insufficiently distributed to support the contagious spread of fire, such as stumps and cones. However, most biomass material can combust and burn, especially under severe weather conditions (severe drought with high winds), so the "biomass is fuel" association seems appropriate for this book. There is often confusion between the singular and plural of fuel; in this book, the term *fuels* is used to describe all the different types and kinds of biomass in the aggregate, while *fuel* is used when referring to one particular type or kind of biomass.

Wildland fuels are the most important environmental factor in fire management. Brown and Davis (1973) mention that "fire ignition, spread, and intensity depend on fuel more than any other factor and it is the fuel that generates the fire behavior with which fire fighters must cope." Scott et al. (2014) say it more simply: "If there is no fuel, there is no fire." Countryman (1969) emphasized that "fuel is the only factor in the fire environment that humans can control." The importance of wildland fuels to fire management cannot be understated and the first step towards fully understanding fuels is learning the basic terminology used in this book.

R. E. Keane, *Wildland Fuel Fundamentals and Applications*,
DOI 10.1007/978-3-319-09015-3_1

1.2 Basic Terminology

1.2.1 Basic Fuel Science Terms

In this book, the *fuelbed* is a general term for the complex array of biomass types for a given area and it is the coarsest scale of fuel description (Fig. 1.1). A comprehensive description of the fuelbed ultimately depends upon the spatial scale. Fuelbeds in forested ecosystems, for example, are somewhat larger than fuelbeds in nonforest ecosystems because the sizes of the trees dictate the scale of canopy fuels. Here, a spatial scale of about 100 m^2 is used to bound or describe a fuelbed regardless of the ecosystem (Fig. 1.1; Sullivan 2009a). This is somewhat greater than the scale of surface fire spread (1–2 m) but is more or less representative of vegetation dynamics (Hiers et al. 2009). Fuelbeds include all types of biomass types and their distributions, and, in this book, they have no vertical height limit, although many have used the term to specifically describe surface fuels. Fuelbeds have specific properties, such as composition, depth, and bulk density, that are used in both fire behavior prediction and fuel management (Chap. 2).

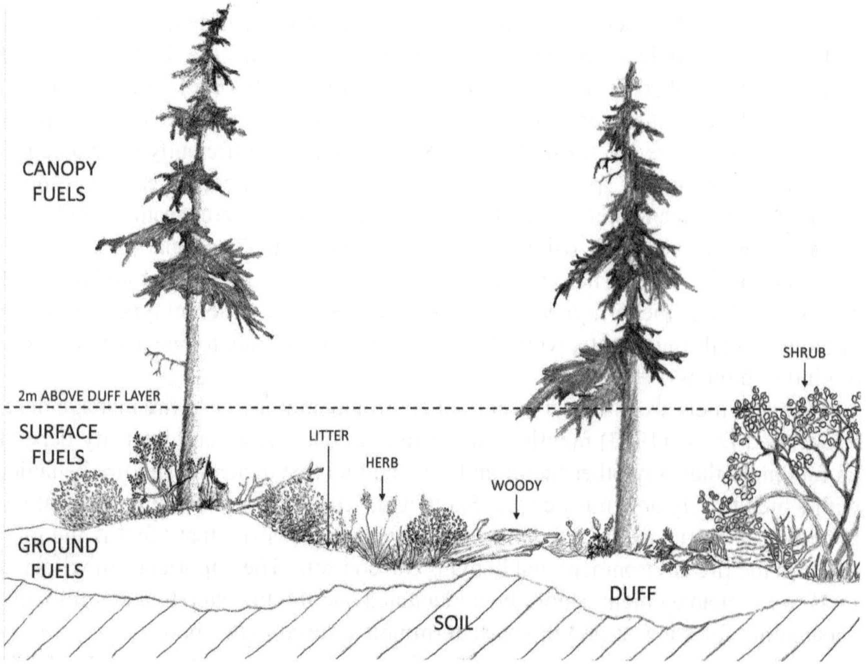

Fig. 1.1 The elements of a typical wildland fuelbed. The full representation of fuels within an area is called a fuelbed. Within a fuelbed, there are three fuel layers: ground, surface, and canopy. Each layer is composed of fuel types, such as litter, shrubs, grasses, and woody biomass in the surface fuel layer

The fuelbed is vertically stratified into three fuel *layers—ground, surface, and canopy* fuels. In this book, *surface fuels* are all biomass within 2 m above the ground surface (Fig. 1.1). This 2-m boundary is mostly arbitrary and was originally defined by several heights depending on the fire application; Brown and Davis (1973), for example, used a 4-ft height. *Ground fuels* are all organic matter below the ground line. The position of the ground line is highly contentious; some put it below the litter at the top of the duff because they feel that only the litter contributes to the propagation of the flaming front, while others put the ground line below the duff because it is incredibly difficult to distinguish between litter and duff in the field. In this book, litter is considered surface fuel while duff is considered ground fuel (Chap. 2). *Canopy fuels* are the biomass above the surface fuel layer. Some define the canopy as starting at 6 m (20 ft; NWCG 2006), while others define it as all tree biomass no matter the height. To be consistent, canopy fuels are defined as all biomass (e.g., shrub, moss, lichen, vine, dead material, and tree) that is higher than 2 m above the ground surface (Fig. 1.1). The term *aerial* fuel is also used to describe canopy fuel (Brown and Davis 1973).

Fuelbed layers are composed of finer-scale elements called fuel types and components (Fig. 1.2). *Fuel types* are general descriptions of the kinds of fuels comprising

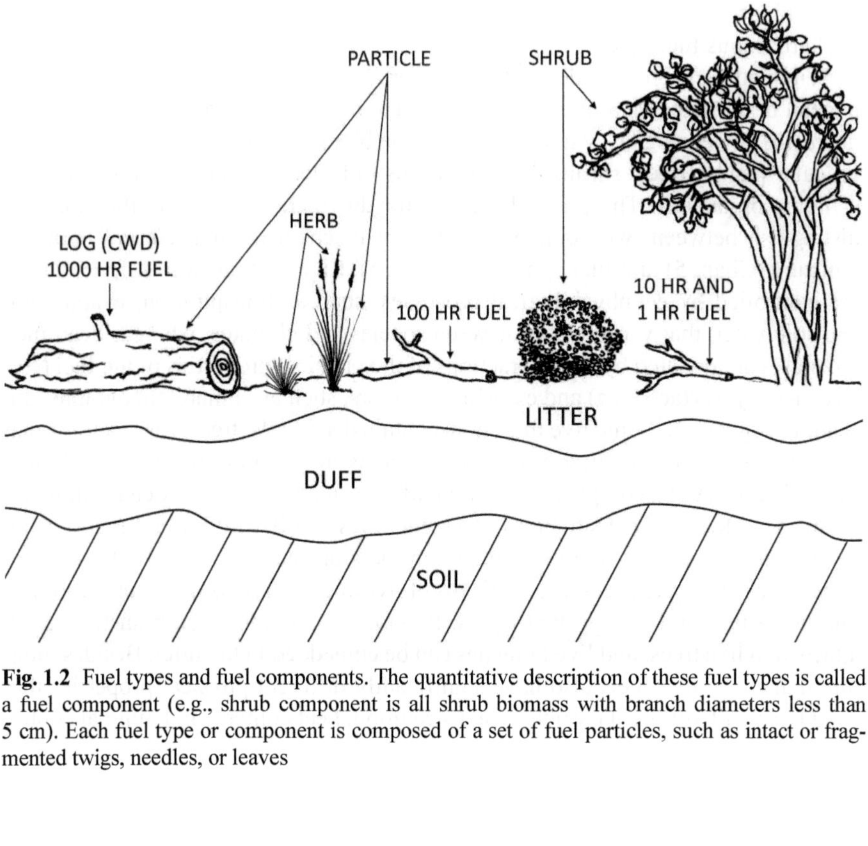

Fig. 1.2 Fuel types and fuel components. The quantitative description of these fuel types is called a fuel component (e.g., shrub component is all shrub biomass with branch diameters less than 5 cm). Each fuel type or component is composed of a set of fuel particles, such as intact or fragmented twigs, needles, or leaves

the fuelbed, whereas *fuel components* are fuel types that are qualitatively and quantitatively defined for specific purposes, mostly for fire behavior prediction. A fuel type might be "woody fuel," while a woody fuel component might be defined as woody fuel of a certain diameter size range (Chap. 3). Many fire practitioners refer to a fuel type as a general term for the dominant fuel of a fuelbed, such as a shrub fuel type describing a fuelbed where the loading is mainly shrubs. In this book, however, a fuel type is specific to the kind of fuel in a fuelbed independent of its loading, and the dominant fuel of a fuelbed is called a *fuel complex* (Bebi et al. 2003). A shrub fuel type would indicate that a fuelbed has some shrub biomass, while a shrub fuel complex would refer to a fuelbed that is dominated by shrubs. Similar to fuelbeds, fuel types and components also have specific properties, such as bulk density, loading (mass per area), and surface area, which are important inputs to fire behavior and effects models and important descriptors of fuel characteristics.

The finest scale of fuelbed description is the fuel *particle,* which is a general term that defines a specific piece of fuel that is part of a fuel type or component of a fuelbed (Fig. 1.2). For example, a fuel particle can be an intact or fragmented stick, grass blade, shrub leaf, or pine needle. Fuel particles have the widest diversity of properties, such as specific gravity, heat content, and shape (Chap. 2), and the properties of fuel components and fuelbeds are often quantified from statistical summaries of the properties of the particles that comprise them. For example, heat content of the herbaceous fuel component may be quantified by averaged heat content estimates across all particles (leaf blades) from all plant species that compose the herbaceous fuel type.

Wildland fuels are also defined as *dead* or *live* within any given fuel type, component, or particle. *Dead* fuel is suspended and downed dead biomass, often called *necromass* by ecologists, while, *live* fuel is the biomass of living organisms, mostly vascular plants (trees, shrubs, herbs), but also of mosses, lichens, and many other living organisms. The principle reason for this dichotomous stratification is to distinguish between two completely different mechanisms that control both fuel moisture (Chap. 5) and fuel dynamics (Chap. 6). Live fuel moistures, for example, are controlled by ecophysiological processes, such as transpiration, evaporation, and soil water, that vary greatly between species and climates, whereas dead fuels moistures are dictated by the interactions of the physical properties of the fuel (e.g., size, density, surface area) and exogenous factors, such as climate, topography, and shading vegetation. Some live fuels may contain dead fuels; trees, for example, may have live wood surrounding dead wood, such as in a healing fire scar. And, most fuelbeds consist of a complex distribution of live and dead fuels so determining live versus dead fuel in field situations can sometimes be difficult, often because some dead fuels may appear as live or they may be attached to live fuels. Mosses and lichens, for example, occur as complexes of live and dead fuels distributed throughout the surface and canopy fuel layers. In another example, dead branches can be attached to live trees, and live branches can be embedded in the litter. Besides moisture dynamics, live fuels also have significantly different physical properties than dead fuels with different particle size distribution, heat content, and mineral content (Chap. 3).

1.2.2 General Fuel Descriptions

Many general terms are also used to describe characteristics of a fuelbed. *Ladder fuel* is a term used to describe a vertical continuous layer of fuel that, when burned, can transport a surface fire into tree crowns to become a crown fire. Ladder fuel can be any live or dead biomass but most of the time it is tree and shrub foliage and small branches that extend down into the surface fuel layer or extend upwards into the canopy layer. *Flashy fuel* is a term often used to describe the finest fuel types that are most easily ignited and combust quickly, such as grasses, small twigs, and litter. *Sound* and *rotten* are two terms that are used to stratify the degree of decay in woody fuel particles although there isn't really a metric that can be used to determine what is a sound log and what is a rotten log. Sound woody fuels usually have greater particle densities, higher heat contents, and lower surface area to volume ratios than rotten fuels (Chap. 3).

Fuel *availability* is a term often used to describe the potential for biomass to burn. For fuel to be available, it must be dry enough to ignite and there must be enough of it to burn. In fact, some older fuel studies have often referred to wildland fuel as only the biomass that is available to burn. A similar term is fuel *condition* often defined as relative flammability of fuel based on type, environment, and mostly fuel moisture (NWCG 2006). Fuel *flammability* is defined as the relative ease at which fuels will burn regardless of amount (NWCG 2006; see Chap. 10). The problem with these three terms is that they are ambiguous, scale-dependent, and difficult to quantify, and therefore, they are often only used to qualitatively describe fuel types and fuelbeds. Some examples of the problems with these terms are illustrated with these questions: Is the fuel available if it only burns in smoldering combustion? Does fuel condition include continuity? Are fuels highly flammable if they produce high intensities? This ambiguity is partially a result of dynamic and complex ecological entities (biomass) being described in a combustion science context.

Fuelbeds are often given names that describe the factors involved in their creation. *Natural* fuelbeds are those fuelbeds created by vegetation development in the absence of disturbance. Endemic (within stand) disturbances may act on the vegetation to also create natural fuelbeds. However, the term "natural" is somewhat ambiguous and open to wide interpretation, so, in this book, the term *undisturbed* fuelbed is used to describe fuelbeds that haven't been affected by major disturbances. Major exogenous disturbances often create their own unique set of fuelbeds that are usually named for the disturbance that created them. *Activity* fuels are those fuelbeds that have been altered by mechanical treatments, such as thinning, timber harvest, and mastication (Hirsch et al. 1979; Fig. 1.3a). The cut or fallen woody fuel particles, such as limbed branches, destroyed tree seedlings, and abandoned tree tops, that are left on the ground after fuel treatments are often referred to as *slash* in activity fuels. *Blowdown* fuelbeds are created by localized high-wind events that topple trees en masse (Woodall and Nagel 2007), while *hurricane* fuelbeds are caused by regional storm events (Busing et al. 2009; Fig. 1.3b). Insect and disease outbreaks often create fuelbeds that many consider hazardous (Jenkins et al.

Fig. 1.3 Examples of various fuelbeds named for the disturbances or vegetation that created them: **a** activity fuelbed (fuel treatment unit in northwestern Montana, USA), **b** hurricane (Shortleaf Pine, Texas USA; courtesy of the Fire and Environmental Rsearch Team, US Forest Service), **c** mountain pine beetle (lodgepole pine central Idaho), and **d** maquis fuel (Sardinia, Italy; photo courtesy of Valentina Bacciu)

2008). Mountain pine beetles, for example, alter both canopy and surface fuels, by killing pines and facilitating canopy growth of surviving competitors (Chap. 6; Fig. 1.3c). However, most fuelbeds are named for those vegetation types that created them; shrub fuel complexes, for example, are sometimes called brush, scrub, maquis, heathlands, and chaparral fuelbeds, depending on the geographical area (Dimitrakopoulos 2002; Keeley et al. 2008; Fig. 1.3d, Chap. 3).

1.2.3 Wildland Fire

Several fire science terms must also be defined to avoid ambiguity when describing fuels (Chap. 2). *Fire behavior* is a general term used to describe physical aspects of the combustion process such as speed and direction of fire spread. Other definitions of fire behavior include "the manner in which fire interacts with topography, fuels, and weather" (NWCG 2006). *Fire spread* is how fast a fire moves in a given direction, while *fire growth* is how large the fire gets in an area over time. *Fire intensity* is the combustion energy released from the burning of organic matter during a fire and is usually described using different metrics, such as reaction intensity, fireline intensity, and radiant energy (Keeley 2009). Fuel types that foster rapid fire spread (fine, flashy fuels) are quite different from those that create hot, intense fires (logs, deep duff, canopy foliage). *Fire effects* are the physical, biological, and ecological impacts of fire on the environment; *fire ecology* is the study of those impacts on living organisms and their environment; and the term *fire severity* is often used to describe the magnitude of fire impacts on the ecosystem (Morgan et al. 2014). *Fire danger* is a term used to describe the combination of both constant and variable factors on fire behavior, such as fuels, weather, and topography, which affect the initiation, spread, and difficulty of control of wildfires.

Wildland fires are usually divided into three types based on the fuel layer in which they are burning. A *crown* fire is the combustion of the canopy fuels above the surface fuel layer, and similarly, a *surface* fire is a fire burning the surface fuels (Chap. 2). A *ground* fire slowly burns the duff and soil organic matter through smoldering combustion. Most wildland fires have all three of these fire types at the same time.

Fires are also described in terms of their effects (Morgan et al. 2014). A *nonlethal surface fire* burns in the surface fuel layer and usually doesn't kill the majority of plants (<20 % mortality), while a *lethal surface fire* results in high plant mortality. A *stand-replacement fire* kills most plants, especially trees in the burned area (>70 % tree mortality; Agee 1993; Morgan et al. 2001). In *mixed severity fires* that contain evidence of the gradient between the other two types of fires distributed across space (Arno et al. 2000).

1.2.4 Modeling

Most fuelbed characteristics are described in terms of input requirements of the fire models that predict fire behavior and effects, so it is important to know the differences between model designs and approaches. This terminology is more descriptive than categorical so it is possible that models can be described by combinations of these terms. An *empirical* model is one that is based on observation and experiment and not on theory. Empiricism forms the basis for much of current fire and fuels research and generally provides the reference against which theory is tested (Sullivan 2009b). Empirical models are often composed of statistical correlations using data measured in the field or derived from laboratory experiments. Some use the term *phenomenological models* when taking a statistical modeling approach because they use information about how a system has typically behaved in the past to develop predictive equations and algorithms; the outcome of the process is predicted using surrogates for the causal mechanisms. Others use the term *statistical* models to indicate that statistics were used to develop the empirical model.

Theoretical models are generated from physical laws, such as those that govern fluid mechanics, combustion and heat transfer. Validation of these kinds of models is extremely difficult, although they may be extrapolated to a wide variety of fire situations. These models may also be called *mechanistic* or *process* models because they take a reductionist approach by explicitly representing the mechanisms that lead from cause to effect. For example, tree growth may be simulated using the ecophysiological processes of photosynthesis and respiration in a mechanistic approach. *Semiempirical* or *quasi-empirical* models are terms used for models that were developed using theoretical equations, but these equations were parameterized using empirical techniques (Sullivan 2009b).

Four key modeling terms are important tasks in fire simulations. *Initialization* is the process of inputting initial parameters into the model. These starting conditions are usually the quantification of fuel properties, such as loading and moisture. *Parameterization* is a term used to describe how the parameters in various equations and algorithms in the model are quantified. Some fuel attributes, such as mineral content, are not dynamic variables that users can input into fire models, but instead they are static parameters that the user cannot modify so it is important to know how these parameters were estimated. Statistical techniques are used for most parameter quantification, but some parameters in theoretical or physical equations can be estimated from the literature. *Calibration* is a term used to describe the adjustment of model inputs and parameters to achieve realistic results. There is always error in the quantification of both parameters and initial conditions in most models, and this error is often reduced by adjusting parameters or inputs so that subsequent simulations produce believable results. And last, *validation* is the process of describing the accuracy and precision of model results. Validation relies on comprehensive databases to use as reference for comparison against simulation results. Every modeling project should involve each of these four phases.

1.3 An Abridged History of Wildland Fuel Science

To fully understand why wildland fuels are described and defined the way there are today, it is important to trace the history of the application of fuels in fire management. Historically, most fuelbeds were described using terms that related more to fire behavior than ecology. Starting in 1919, Show and Kotok (1930) correlated fire behavior and firefighting descriptors to vegetation cover types to represent fuels, and called categories in this classification "hour control zones," which represented the time it took for a suppression force to arrive after an ignition. Hornby (1935) described fuels of the northern Rockies as categories in an ordinal fire behavior classification that integrated resistance to spread and suppression effectiveness levels. This approach was then employed to describe and map fuels for many other areas of the USA including the mid-west (Jemison and Keetch 1942), the mountains and seaboard of the Atlantic region (Banks and Frayer 1966), the Pacific Northwest (Abell 1937), and parts of New Jersey (New Jersey Department of Conservation and Development and US Department of Agriculture 1942). Both Barrows (1951) and Banks and Frayer (1966) revised the Hornby (1936) methods to include a fuel classification key that integrated fire behavior categories with vegetation and structural characteristics. Matthews (1937) even developed a plot-based sampling method to sample these fire behavior categories to describe fuel at finer scales. These fuel classification and mapping efforts had many problems, mostly because they described fire behavior not fuels. Brown and Davis (1973) recognized several other reasons why fuel descriptions based on fire behavior were ineffective: (1) expensive (costly to train and implement), (2) lack of detail (too broad to be applied locally), (3) obsolescence (mapped fuel types rapidly changed over a short time), (4) narrowly focused (evaluated for worst case burning conditions and envisioned the area burning in only large fires), (5) limited application (could not be used for other fire management tasks), and most importantly, (6) no associated comprehensive technique for measuring fuels.

Another historical approach often used for assessing fuels involved naming and describing fuelbeds based on vegetation characteristics. Mitchell (1929), for example, described the unique fuels of the mid-western USA using vegetation types. Fuel types in New Jersey, USA, were named after forest vegetation types for fire danger prediction (Little 1945). The basis of the Show and Kotok (1930) fuel descriptions was broadly defined vegetation types. Barrows (1951) stratified fire occurrence statistics by two vegetation-based fuel types (timber and grass), three management activity types (cutover, burned, forested), and several forest types in his description of wildfires in the US northern Rocky Mountains. Wendel et al. (1962) sampled fuel weights for various vegetation types in southeastern USA and then used the fuel weights to assign potential fire behavior ratings. Fuel classification systems for Ontario and New Brunswick, Canada, were based on vegetation characteristics (Walker 1971).

Both of these historical approaches ignored the inherent complexity of a fuelbed and attempted to simplify fuelbed descriptions into something that could be easily understood by managers. It was much easier to relate a fuelbed to a recognizable vegetation type or to some abstract interpretation of fire behavior than directly quantify the diverse array of fuel types in a fuelbed. The main reason for this was simple; there really wasn't any reason to stratify the fuelbed into its components. It wasn't until analytical tools, methods, and models were developed for fire management that there became a reason for dissecting fuelbeds into components and describing component properties.

The prediction of fire danger was the first concerted effort at creating a fire management tool (Hardy and Hardy 2007). Gisborne (1936), for example, differentiated fuel types in the fuelbed to more accurately estimate fuel moisture to predict fire danger and Curry and Fons (1938) differentiated fuel types to predict fire spread for fire danger. Fahnestock (1970) developed one of the first comprehensive fuel assessment methods that described the fuelbed as a complex of integrated fuel types. He used various fuel type properties, such as size, shape, and continuity, of three different fuel layers (ground, surface, crown) to rate the potential for spread and crowning. In the 1960s and 1970s, fire scientists around the world started creating fire behavior models that were then implemented into a variety of fire behavior prediction systems for managers. These systems required users to input specific fuel information by component and property (Rothermel 1972; McArthur 1966). Fire managers now had a quantitative description of a fuelbed that had a direct application—the simulation of fire behavior and effects. While there have been many modifications to fuel descriptions since 1960, wildland fuelbeds have mostly been described using a suite of components and properties that were specifically engineered for fire behavior computations.

The main problem with this engineering approach is that it is often incompatible with describing the dynamic ecology of wildland fuelbeds. Woody fuels, for example, may be defined by particle diameter classes with ranges that are so broad that the variability of biomass estimates within a diameter class may overwhelm differences across fuelbeds. Rates of decomposition and deposition of woody fuel particles may also vary greatly over the diameters of particles within one class. And because fire behavior-engineered components often tend to have high variabilities in the properties that are used to define them, it may be more difficult to quantify and evaluate important fire management concerns, such as fuel treatment longevity and effectiveness. Additionally, it may be more difficult to get accurate estimates of other fire-related management issues, such as smoke emissions, tree mortality, and fuel consumption, when high variability is a result of inappropriate fuel descriptions. Because wildland fuel science now has a much broader application than just fire behavior, such as fire effects, wildlife habitat assessment, carbon inventory, and tree regeneration potential, it may be time to take a more ecological approach to studying fuels.

References

Abell CA (1937) Rate of spread and resistance to control data for Region 7 fuel types and their application to determine strength and speed of attack needed. Preliminary Report. U.S. Department of Agriculture, Forest Service, Asheville, 7 pp

Agee JK (1993) Fire ecology of Pacific Northwest forests. Island Press, Washington

Arno SF, Parsons DJ, Keane RE (2000) Mixed-severity fire regimes in the northern Rocky Mountains: consequences of fire exclusion and options for the future. In: Wilderness science in a time of change conference, volume 5: Wilderness ecosystems, threat, and management, Missoula, Montana, May 23–27, 1999. Fort Collins, CO. U.S. Dept. of Agriculture Forest Service Rocky Mountain Research Station, Proceedings RMRS-P-15-Vol 5, pp 225–232

Banks WG, Frayer HC (1966) Rate of forest fire spread and resistance to control in the fuel types of the eastern region. U.S. Department of Agriculture, Forest Service, Washington, DC. Fire Control Notes 27(2)10–13

Barrows JS (1951) Fire behavior in northern Rocky Mountain forests. Northern Rocky Mountain Forest & Range Experiment Station Paper 29, 274 pp

Bebi P, Kulakowski D, Veblen TT (2003) Interactions between fire and spruce beetles in a subalpine Rocky Mountain forest landscape. Ecology 84(2):362–371. doi:10.1890/0012-9658(2003)084[0362:ibfasb]2.0.co;2

Brown AA, Davis K (1973) Forest fire control and use, 2nd edn. McGraw-Hill, New York

Busing RT, White RD, Harmon ME, White PS (2009) Hurricane disturbance in a temperate deciduous forest: patch dynamics, tree mortality, and coarse woody detritus. Plant Ecol 201(1):351–363. doi:10.1007/s11258-008-9520-0

Countryman CM (1969) Fuel evaluation for fire control and fire use. Paper presented at the Symposium on fire ecology and control and use of fire in wildland management, Tucson, AZ
Curry JR, Fons WL (1938) Rate of spread of surface fires in the Ponderosa pine type of California. J Agric Res 57(4):239–267
Dimitrakopoulos AP (2002) Mediterranean fuel models and potential fire behaviour in Greece. Int J Wildland Fire 11(2):127–130
Fahnestock GB (1970) Two keys for appraising forest fire fuels. USDA Forest Service. Pacific Northwest Forest and Range Experiment Station, Portland
Gisborne HT (1936) The principles of measuring forest fire danger. J For 34(8):786–793
Hardy CC, Hardy CE (2007) Fire danger rating in the United States of America: an evolution since 1916. Int J Wildland Fire 16(2):217–231
Hiers JK, O'Brien JJ, Mitchell RJ, Grego JM, Loudermilk EL (2009) The wildland fuel cell concept: an approach to characterize fine-scale variation in fuels and fire in frequently burned longleaf pine forests. Int J Wildland Fire 18(3):315–325. doi:10.1071/WF08084
Hirsch SN, Meyer GF, Radloff DL (1979) Choosing an activity fuel treatment for southwest ponderosa pine. USDA Forest Service. Rocky Mountain Forest and Range Experiment Station, Fort Collins
Hornby LG (1935) Fuel type mapping in Region One. J For 33(1):67–72
Hornby LG (1936) Fire control planning in the northern Rocky Mountain region. USA Report Number 1. USDA Forest Service. Northern Rocky Mountain Forest and Range Experiment Station, Missoula, 18 pp
Jamison GM, Keetch JJ (1942) Rate of spread of fire and its resistance to control in fuel types of Eastern Mountain Forests. Appalachian Forest Exp. Sta. Tech. Note No. 52. 44 pp
Jenkins MA, Hebertson E, Page W, Jorgensen CA (2008) Bark beetles, fuels, fire and implications for forest management in the Intermountain West. For Ecol Manag 254:16–34
Keeley JE (2009) Fire intensity, fire severity and burn severity: a brief review and suggested usage. Int J Wildland Fire 18(1):116–126.
Keeley JE, Brennan T, Pfaff AH (2008) Fire severity and ecosytem responses following crown fires in California shrublands. Ecol Appl 18(6):1530–1546
Little S (1945) Influence of fuel types on fire danger. J For 43(10):744–749
Matthews DN (1937) Small-plot method of rating forest fuels. J For 35(10):929–931
McArthur AG (1966) Forest fire danger meter. Commonwealth of Australia, Forestry and Timber Bureau, Canberra, Australia
Mitchell JA (1929) Forest fire hazard as affected by weather conditions, forest type, and density of cover. Research Bulletin. U.S. Department of Agriculture, Forest Service, Madison
Morgan P, Hardy CC, Swetnam TW, Rollins MG, Long DG (2001) Mapping fire regimes across time and space: understanding coarse and fine-scale fire patterns. Int J Wildland Fire 10:329–342
Morgan, P, Keane, RE, Dillon G, Sikkink P Silverstein, R, Jain, T, Hudak, A, Strand, E (2014) Challenges with fire and burn severity and its application in fire management. Int J Wildland Fire (in press)
New Jersey Department of Conservation and Development, U.S. Department of Agriculture FS (1942) Forest fuel types of New Jersey. New Jersey Department of Conservation and Development, Trenton, NJ
NWCG (2006) National Wildfire Coordination Group Glossary of Fire Terminology. http://www.nwcg.gov/pms/pubs/glossary/
Rothermel RC (1972) A mathematical model for predicting fire spread in wildland fuels. United States Department of Agriculture, Forest Service, Intermountain Forest and Range Experiment Station, Ogden, Utah Research Paper INT-115, 88 pp
Scott AC, Bowman DMJS, Bond WJ, Pyne SJ, Alexander ME (eds) (2014) Fire on earth: an introduction. Wiley, Chichester
Show SB, Kotok E (1930) The determination of hour control for adequate fire protection in the major cover types of the California pine region. US Department of Agriculture Technical Bulletin 209, 33 pp

Sullivan AL (2009a) Wildland surface fire spread modelling, 1990–2007. 1: Physical and quasi-physical models. Int J Wildland Fire 18(4):349–368

Sullivan AL (2009b) Wildland surface fire spread modelling, 1990–2007. 2: Empirical and quasi-empirical models. Int J Wildland Fire 18(4):369–386

Walker JD (1971) A simple subjective fuel classification system for Ontario. Information Report O-X-155. Environment Canada, Forestry Service, Sault Ste. Marie, Ontario, Canada, 7 pp

Wendel GW, Storey TG, Byram GM (1962) Forest fuels on organic and associated soils in the coastal plain of North Carolina. Research Report No. 144. US Department of Agriculture, Forest Service, Asheville, NC, 23 pp

Woodall CW, Nagel LM (2007) Downed woody fuel loading dynamics of a large-scale blowdown in northern Minnesota, USA. For Ecol Manage 247(1–3):194–199

Chapter 2
Fundamentals

Fire's the sun, unwindin' itself out o' the wood
David Mitchell, author

2.1 Fire and Fuel Basics

2.1.1 Fuel Chemistry

Wildland fuels are mostly created from plants, sunlight, water, and nutrients to be eventually burned by fire. Fuels are created by plants as a product of photosynthesis, a chemical process where carbon dioxide (CO_2), water (H_2O), and energy from the sun (solar radiation) are used to produce organic compounds of the chemical form $(C_6H_{10}O_5)_y$ and also oxygen (O_2). This can be expressed in the general formula:

$$5H_2O + 6CO_2 + \text{Solar Energy} \rightarrow (C_6H_{10}O_5)_y + 6O_2. \tag{2.1}$$

The substances that compose biomass $(C_6H_{10}O_5)_y$ are quite susceptible to burn because of their organic chemical constituency. The primary substances found in plant biomass are cellulose, hemicellulose, and lignin. In general, wood, an important fuel in forest ecosystems, is composed of around 40–55 % cellulose, 15–25 % hemicellulose, 15–30 % lignin, and 2–15 % other matter, while needles have less lignin but more cellulose than wood.

Gisborne (1947) said "all fuels have pretty much the same chemical constituents (cellulose, starch, and lignin) and when these organic fuels burn in a wildland fire, they combine with oxygen to create carbon dioxide, water, and heat" as denoted in the following formula:

$$(C_6H_{10}O_5)_y + 6O_2 \text{ heat} \rightarrow 5H_2O + 6CO_2 + \text{Heat} + \text{secondary compounds} \tag{2.2}$$

R. E. Keane, *Wildland Fuel Fundamentals and Applications*,
DOI 10.1007/978-3-319-09015-3_2

This is a chemical representation of the process of combustion, often considered a chemical chain reaction because the heat produced by *combustion* acts as a catalyst which further increases the rate of reaction. Byram (1959) presented the following chemically balanced oxidation reaction for complete combustion of plant biomass:

$$4(C_6H_9O_4)+25O_2+[0.322M\ H_2O+94N_2]\rightarrow 18H_2O+24CO_2$$
$$+[0.322M\ H_2O+94N_2]+[11.6\times 10^9\ \text{J Heat}] \quad (2.3)$$

Note that 4 kg moles of plant material ($C_6H_9O_4$) yields about 11.6 billion joules of heat, and dividing this heat by the mass gives the heat content of the fuel (see Sect. 2.3.1.1).

2.1.2 Scales of Combustion

The physical process of combustion is quite complex and occurs in at least four overlapping phases (Zhou and Mahalingam 2001; Bebi et al. 2003). In the *pre-ignition phase,* unburned fuel ahead of the advancing flame front is heated and raised to its ignition temperature in a series of endothermic (requiring heat) reactions dominated by dehydration and the volatilization of organics. Water is vaporized in the cell structure, then driven to the surface of the fuels, and vented to the atmosphere. The second phase of combustion known as *pyrolysis* begins as fuel temperature rises and cellulose and other compounds begin to decompose to release combustible organic gases and vapors, thereby converting biomass into volatiles, tars, char, and ash. Cellulose is pyrolyzed between 280 and 400 °C through dehydration and depolymerization, while lignin is pyrolyzed at temperatures of 280–500 °C because it is more complex and thermally stable (Liodakis et al. 2002). The *combustion phase* occurs when the burning process becomes exothermic (generating heat) in the presence of oxygen giving off energy in the form of heat and light, and the start of combustion is often termed ignition. Flaming combustion occurs when volatized gases are oxidized and flames are generated, usually occurring when the temperature of the volatiles reach 450–500 °C. Combustion without flames is called smoldering combustion, which is the surface oxidation of char, which provides just enough heat to continue pyrolysis. In general, the *smoldering combustion phase* occurs when the concentration of combustible vapors above the fuel is too small to support a persistent flame, so gases and vapors condense, appearing as smoke. Once most volatile gases have been driven off, the *glowing combustion phase* occurs, where only embers and smoke are visible and there is little smoke; the carbon remaining in the fuel is oxidized to continue to produce significant heat.

This complex combustion process is often simplified so that it can be taught to fire specialists using the famous "fire triangle" (Fig. 2.1). At the finest spatial scale, a combination of three elements is needed for a wildland fire to burn: heat, oxygen, and fuel (Countryman 1969). The heat source for ignition can be from lightning, matches, drip torches, or, mostly, the fire itself. Oxygen is in great supply in the earth's atmosphere, but sometimes the combustion process itself may use

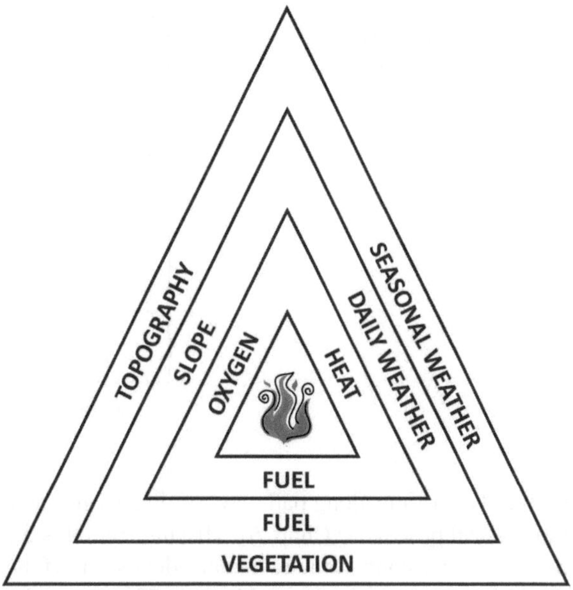

Fig. 2.1 A new variation on the traditional fire triangle often used to teach fire science to managers. The *inner triangle* refers to combustion at the flame level, the *middle triangle* refers to fire spread at a stand level, and the *outside triangle* refers to fire growth at the landscape level

more oxygen than can be supplied by the atmosphere, thereby governing burning rates. And last, there is fuel. As Van Wagner (1983) mentions, the fuel must be the appropriate size and arrangement to facilitate fire spread and it must be dry enough for combustion (i.e., low moisture content). Unfortunately, the inner fire triangle in Fig. 2.1 really only works at very small scales; perhaps the scale of the flame, which may be useful for firefighters but somewhat ineffectual for the diverse and complex issues facing a wildland fuel manager. Therefore, many have added additional fire triangles to represent the scaling of combustion to a fire event (Fig. 2.1; Alexander 2014).

However, to fully understand fuels, it is important to recognize that the process of combustion scales from the flame to burning period to fire event over various time and space scales (Fig. 2.1). A more comprehensive representation of the fire triangle is detailed by Moritz et al. (2005; Fig. 2.2) where fire moves across an area and interacts with topography (slope, aspect), weather (temperature, humidity), and the fuel complex. At coarser scales, the fuel properties important to fire spread are governed more by the distribution of fuels across the landscape or contagion (continuity of a fuelbed). The landscape-level spatial scale of fire spread best describes the operational management of fuels and is probably the most appropriate for designing fuel treatments (Agee and Skinner 2005). However, some large fires can burn entire landscapes over the course of weeks, and as more fires burn the same landscape over hundreds of years, these fires interact with previous fires, climate (drought, warming), ignition patterns (lightning, humans), and vegetation to create a *fire regime* (Chap. 6). In Fig. 2.2, fuels are represented by vegetation to signify that fuel conditions change over time and this change is mediated by vegetation development processes (regeneration, growth, mortality) and succession (species

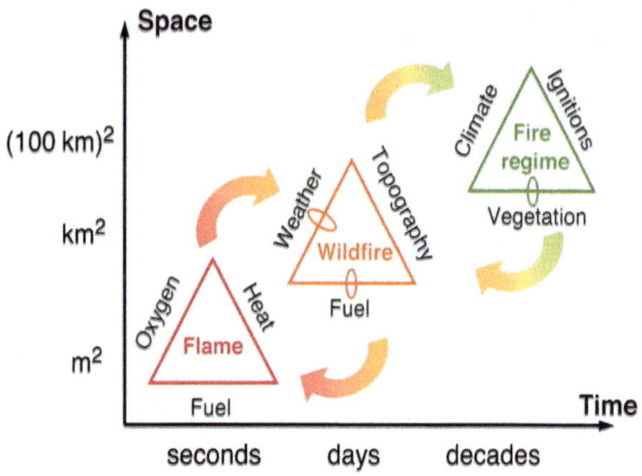

Fig. 2.2 The scaling of the combustion process over time and space from Moritz et al. (2005)

replaced over time along pathways of disturbance adaptations, shade tolerance, and biophysical processes; Chap. 6). Missing from this diagram are the key biophysical processes that control fuels, namely deposition (fallen plant material often called litterfall) and decomposition (Chap. 6). The interactions of plant succession and endogenous and exogenous disturbances with biomass deposition and decomposition mostly govern fuel properties, fuelbed dynamics, and spatial distributions at local to landscape scales.

This brings up an interesting dilemma in that most fire behavior research has been done at the fuelbed or flame scale (Sullivan 2009a) often resulting in a scale mismatch between fuel management and fire behavior, because most fuel management issues demand a coarser scale of analysis (Keane et al. 2012a). Therefore, an overview of how fuels are defined in surface fire behavior models is needed to understand the current and past use of fuels.

2.2 Surface Fire Behavior Modeling

2.2.1 Fire Behavior Formulation

The great fires of 1910 created the first real need for an understanding of fire behavior and search for models to predict fire behaviors (Pyne 2001). US fire pioneers, such as Gisborne (1927) and Hawley (1926), linked empirical evidence with observed fire characteristics to explain the behavior of fire. Later, Curry and Fons (1938) and Fons (1946) attempted to describe fire spread using more theoretical, physically based relationships. However, it quickly became evident to fire managers that these physical relationships were too complex to easily apply on the fire line. Fire managers needed some way to easily estimate fire behavior to more effectively manage wildfires, predict effects of prescribed burns, and save firefighter's lives.

Table 2.1 Fuel characteristics and properties used as input parameters for basic fire behavior equations at the three scales of fuelbed description

Scale	Symbol	Parameter	Notes
Fuel particle	d	Particle diameter	Often stratefied into classes
	SAVR	Surface-area-to-volume ratio (m^2/m^3)	Called SAVR in this book but is also called σ in many fire texts
	ρ_p	Particle density (kg m^{-3})	Generally 500 kg m^{-3} (32 lb ft^{-3})
	FMC	Moisture content (fraction)	Dry weight basis kg moisture per kg wood
	S_e	Effective mineral content (fraction)	Generally 0.010 (kg minerals – kg silica) per kg wood
	S_T	Total mineral content (fraction)	Generally 0.0555 kg minerals per kg wood
	h	Heat content	Often 18586 J kg^{-1} (8000 BTU lb^{-1})
Fuel component	W	Fuel loading (kg m^{-2})	Oven-dried fuel weight; a highly dynamic input
	ρ_b	Component bulk density (kg m^{-3})	Generally an integrated average across the fuel component and includes air space
	M_x	Dead fuel moisture of extinction (fraction)	Live fuel moisture of extinction is not in the basic model; another highly dynamic input
	δ	Surface fuel layer depth (m)	Mean fuelbed value
	β	Surface fuel layer packing ratio (dimensionless)	See Table 2.2 for estimation

Critically needed was an estimate of how fast a fire burned, called rate of spread (R), because this was identified as an important characteristic in firefighter deaths (Barrows 1951). An additional estimate of how hot a fire burned, called fire line intensity (I) or the rate of heat release per unit length of the fire front, was needed to determine when a fire is too hot to fight. Byram (1959) defined fire line intensity, I, as:

$$I = \frac{hW_cS}{60} \tag{2.4}$$

where h is the heat yield of the fuel (kJ kg^{-1}), S is the forward rate of spread of the fire (m min^{-1}), and W_c is the weight of fuel consumed in flaming combustion (kg m^{-2}). The number 60 is a conversion factor so that the units for I are kW m^{-1} (kJ m^{-1} min^{-1}). Fuel weight consumed (W_c) depends on initial fuel loading (W; kg m^{-2} dry weight). Both loading (W) and heat yield of fuel (h, often called heat content) are the first two important fuel properties for predicting fire behavior (Table 2.1). Linking fire intensity (I) with spread rate (R) provided a means to evaluate the potential to suppress the fire using the fire characteristics chart (Andrews and Rothermel 1982).

Rothermel (1972) and his team used results of this previous work to create the quasi-empirical mathematical model that is now integrated into a wide variety of US fire behavior prediction systems, such as BEHAVE (Andrews 2014), FARSITE (Finney 1998), and FIREHARM (Keane et al. 2010). This model has been extensively modified, adjusted, and refined (Albini 1976; Andrews 1986), but the main equation for fire spread prediction still takes the general form (Table 2.2):

Table 2.2 Equations of the basic fire spread model taken from Rothermel (1972) and Albini (1976) (courtesy of Pat Andrews) and converted to metric units

Variable	Equation	Equation number Rothermel (1972)
Rate of spread ($m\ min^{-1}$)	$R = \frac{I_R \xi (1+\phi_W+\phi_S)}{\rho_b \varepsilon Q_{ig}}$	(Eq. 52)
Reaction intensity ($kW\ m^{-2}\ min^{-1}$)	$I_r = \Gamma' w_n h \eta_M \eta_s$	(Eq. 27)
Fire line intensity ($kW\ m^{-1}$ fire line)	$I_b = \frac{12.61_r R}{60(SAVR)}$	Added later
Optimum reaction velocity (min^{-1})	$\Gamma' = \Gamma'_{max} (\beta / \beta_{op})^{[A} e^{A(1-\beta/\beta_{op})]}$, where $A = 133(SAVR)^{-0.7913}$	(Eqs. 38, 39)
Maximum reaction velocity (min^{-1})	$\Gamma'_{\max} = (SAVR)^{1.5}(495+0.0494(SAVR)^{1.5})^{-1}$	(Eq. 36)
Optimum packing ratio (fraction)	$\beta_{op} = 3.348(SAVR)^{-0.8189}$	(Eq. 37)
Packing ratio (fraction)	$\beta = \frac{\rho_b}{\rho_p}$	(Eq. 31)
Oven-dry bulk density ($kg\ m^{-3}$)	$\rho_b = \frac{w_o}{\delta}$	(Eq. 40)
Net fuel loading ($kg\ m^{-2}$)	$w_n = w_o(1-S_T)$	(Eq. 24) replaced by Albini (1976)
Moisture damping coefficient (fraction)	$\eta_M = 1-2.59 r_M + 5.11(r_M)^2 - 3.52(r_M)^3$, where $r_M = FMC / M_x$ (max=1.0)	(Eq. 29)
Mineral damping coefficient (fraction)	$\eta_s = 0.174 S_e^{-0.19}$ (max=1.0)	(Eq. 30)
Propagating flux ratio (fraction)	$\xi = (192+0.2592(SAVR))^{-1e[(0.792+0.681(SAVR)^{0.5})(\beta+0.1)]}$	(Eq. 42)
Wind factor (fraction)	$\phi_w = CU^B(\beta/\beta_{op})^{-E}$, where $C = 7.47e^{(-0.133(SAVR)^{0.55})}$; $B = 0.02526(SAVR)^{0.54}$; $E = 0.715e^{[-3.59\times10^{-0.4}(SAVR)]}$	(Eq. 47–50)
Slope factor (fraction)	$\phi_s = 5.275\beta^{-0.3}(\tan\phi)^2$	(Eq. 51)
Effective heating number	$\varepsilon = e^{(-138/SAVR)}$	(Eq. 14)
Heat of pre-ignition ($kW\ kg^{-1}$)	$Q_{ig} = 250+1116FMC$	(Eq. 12)

$$R = \frac{I_r \xi (1 + \phi_W + \phi_s)}{(\rho_b)(\varepsilon)(Q_{ig})} \tag{2.5}$$

where I_r is reaction intensity, ξ is the propagating flux ratio (dimensionless), φ_w is a scaling function for wind (number between zero and one), φ_s is a scaling function for slope, ρ_b is the bulk density of the fuelbed (kg m^{-3}), ε is the effective heating number, and Q_i is the heat of pre-ignition (kJ kg^{-1}; Table 2.2). Reaction intensity (I_r) can be estimated by the amount of fuel consumed (W_c) and that fuel's heat content (h) using a reformulation of the Andrews and Rothermel (1982) relationship:

$$I_r = \frac{W_c h}{t_r} \tag{2.6}$$

where t_r is the residence time (min) that is computed from the Anderson (1969) empirical relationship:

$$t_r = \frac{12.595}{\text{SAVR}} \tag{2.7}$$

where SAVR is the characteristic surface-area-to-volume ratio (m^{-1}) of the fuelbed. SAVR is the third important fuel property because it is in the majority of fire behavior calculations (Table 2.2). The characteristic SAVR is estimated from the weighted averages across all surface fuel components specified in the model and the SAVR values for each fuel component is estimated as an average for each particle using the diameter in the following equation:

$$\text{SAVR} = \frac{4}{d} \tag{2.8}$$

where d is the average diameter of the particles in the fuel component (m). Particle diameter is the fourth important fuel property because it is related to SAVR and it is used to estimate loading. The next important fuel property is the parameter ρ_b (bulk density of the fuelbed, kg m^{-3}). This parameter is also used to estimate the effective heating number (ε) in Eq. 2.5 using the empirical Rothermel (1972) relationship:

$$\epsilon = \frac{\rho_b}{\rho_e} = e^{\frac{-138}{\text{SAVR}}} \tag{2.9}$$

where ρ_e is the effective fuelbed bulk density (kg m^{-3}; Table 2.2). However, the effective heating number can also be accurately estimated from SAVR, which is used to represent fuel particle size (Eq. 2.8). Fuelbed bulk density (ρ_b) is often calculated from the following equation:

$$\rho_b = \frac{W}{\delta} \tag{2.10}$$

where δ is the fuelbed depth (m), the sixth important fuel property, and W is fuelbed loading (kg m^{-2}).

In the Rothermel (1972) algorithms, reaction intensity (I_r) is computed from another equation (see Table 2.2):

$$I_r = \Gamma' W_n h \eta_m \eta_s \tag{2.11}$$

where W_n is fuel loading (kg m^{-2}) adjusted for the mineral content, Γ' is the reaction velocity (a dynamic variable that represents the rate and completeness of fuel consumption), and η_m and η_s are damping functions to account for the effect of fuel moisture and mineral content, respectively, on combustion (equations for all variables in Table 2.2). Two fuel properties have a major effect on reaction intensity. Increasing fuel moisture and mineral content decreases I_r using the damping coefficients η_m and η_s that are represented by empirical relationships. The coefficient η_m is calculated using an empirical polynomial regression equation where the only variable is the ratio of the fuel moisture content (FMC; %) to the moisture of extinction (M_x; %; Eq. 29 in Rothermel (1972); see Table 2.2). These two fuel moisture variables are the seventh and eighth important fuel property. The mineral content damping coefficient (η_s) is calculated using the following empirical equation developed by Philpot (1970):

$$\eta_s = 0.174 S_e^{-0.19} \tag{2.12}$$

where S_e is the effective mineral content calculated as the amount of silica in the fuel component minus the mineral content (S_T). Mineral content (S_e and S_T) is the ninth important fuel property. Fuelbed compactness is another important fuelbed property affecting I_r and it is often represented by the packing ratio (β) defined by:

$$\beta = \frac{\rho_b}{\rho_p} \tag{2.13}$$

where ρ_p is the average particle density of the particles that comprise the fuel component (kg m^{-3}), the eleventh important fuel property (Table 2.1).

2.2.2 *Fire Behavior Assumptions*

To simplify the spatial complexity of the combustion process, early fire scientists had to make the assumption that fire spread can be represented by the movement of a flame across a semipermeable surface using a one-dimensional point model (Fig. 2.3; Rothermel 1972). This would have been a good assumption if (1) fuels were homogeneously distributed over the scale of a burning, (2) fires act at only one scale, and (3) the scale of the fire matched the scale of the fuels. But unfortunately, the distribution, condition, characteristics, and consumption of burnable biomass are highly complex over space and time (Frandsen and Andrews 1979; Chap. 6). Therefore, the scale mismatch between fire modeling and fuel properties may bias the simulation of fire in a one-dimensional approach. For example, an input fuel parameter that varies greatly over the small scales of fire spread, such as fuelbed

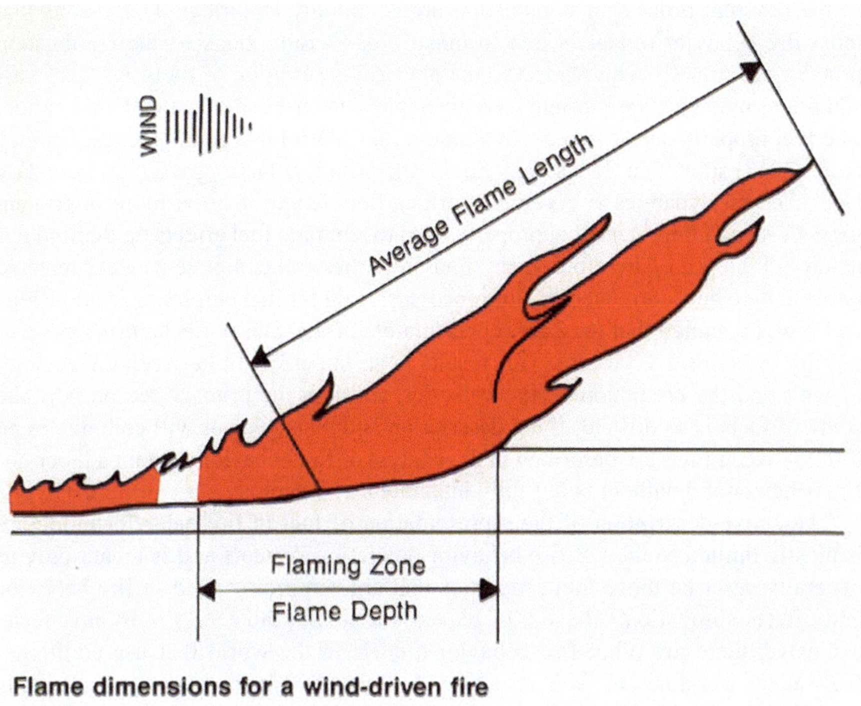

Fig. 2.3 The theoretical flame of fire spread used in the development of the one-dimensional fire spread models

bulk density, may not provide great predictive power when fire is simulated for one point in space.

Another great need for fire behavior simulation was to predict fire effects. Byram (1958) noted the importance of predicting the impact of fire on living vegetation, and Rothermel and Deeming (1980) noted fire behavior was critical for predicting fire effects. Yet ironically, fuel inputs to most fire models were engineered to fit combustion relationships without an ecological context. Successful prediction of fire effects requires that the model to be designed so that the inputs make ecological sense (Keane and Finney 2003) and that the outputs are germane to the assessment of fire effects. Grouping all log biomass into one size class, for example, ignores the great importance of log size on fuel properties and subsequent combustion, and, more importantly, on the effects of burning different-sized logs on soil heating and smoke production. Moreover, some fire effects models use fuel properties that are not used in fire behavior simulation algorithms (Trakhtenbrot et al. 2014). Mechanistic fire-caused tree mortality models, for example, use thermal conductivity to simulate heat flow through bark (Mitchell 2013).

And last, it is important to note that most operational fire behavior models are quasi-empirical in design (Sullivan 2009b) in that they predict fire spread and intensity using physically based statistical algorithms. As Finney et al. (2013) mention, there really is no physical theory of fire spread, so many numerical representations of the physical process of combustion are commonly empirical. This means that most fire behavior modelers had to make broad assumptions of the combustion process, specifically with respect to the physical description of fuels, and these assumptions may be inappropriate for a given physical process or scale of application. The fuel property surface-area-to-volume ratio (SAVR) is a good example. Finney et al. (2013) show that SAVR may not be the principal factor governing boundary layer thermal dynamics and vertical surface flow length may be more important, yet SAVR is an important fuel property used to simulate fuel effects on thermal dynamics (Table 2.2). Therefore, many fuel properties and components were selected because they best correlated to fire processes using limited empirical relationships and it was assumed that they are representative of the causal mechanisms governing fire behavior everywhere. This results in an imperfect fit between the ecology of fuels and the prediction of fire behavior, and it is the primary reason why the study of fuels is so difficult. Fuel description and management will continue to be difficult when fuels are described in the context of fire behavior without a theory of fire behavior and without being fully integrated with ecology.

The above description of the representation of fuel in fire behavior modeling is mostly limited to the US fire behavior prediction systems and is meant only to generally describe those fuel properties that are commonly used in fire behavior and effects simulation. The list of eleven variables (Table 2.1) is by no means exclusive; there are other fire behavior models in the world that use additional fuel-related variables in their structure (Sullivan 2009a, b; Linn 1997; Parsons et al. 2010). Moreover, there are many other fuel particle and fuelbed properties that are important to the field of wildland fuel ecology, such as degree of rot, particle length, and fuelbed cover, that are not discussed here. However, this list (1) probably represents those fuel properties used across most of the world's fire behavior modeling systems, (2) is perhaps the most important for the merging of fire behavior with ecology, and (3) contains properties that can be measured by fire behavior practitioners and wildland fuel managers. These properties are discussed below fuelbed scale.

2.3 Surface Fuel Properties

Past fuel studies have identified the fundamental properties of fuels as quantity, size, shape, arrangement, continuity, and pattern (Bebi et al. 2003; Ottmar et al. 2007), but this classical list has many limitations. First, there are scale inconsistencies, in that some properties refer to individual particles, while others refer to all particles in fuel components, layers, and fuelbeds. Second, missing are some physical properties that describe the role of the fuel in the combustion process, especially in the context of fire behavior (see Sect. 2.2). There is also a missing linkage between many of these fundamental properties and how they are used to simulate fire or how they are employed in fire and fuel management. For example, arrangement remains unaddressed in point-scale fire models. And last, this list is missing critical metrics and variables that can be used to quantify the properties. This chapter discusses the quantity (loading), size, and shape of surface fuels in the following section by scale and specific property. Arrangement, continuity, and pattern are discussed in detail in Chap. 6. While some important fuel parameters listed in Table 2.2 refer to the properties of a fuel component, the component parameter is often computed as an average across particles within the component. Therefore, each fuel property is discussed at the scale of measurement rather than the scale of model input.

2.3.1 Particle Properties

2.3.1.1 Particle Diameter (*d*)

While fuel particle diameter (*d*) is a critical fuel parameter for fire behavior modeling, it is even more important in sampling for fuel loading (*W*). Thomas (1953), for example, mentioned that the duration of burning is related to stick diameter by approximately the 1.5 power. However, use of diameter in most fire and fuel applications may be overgeneralized because nearly all fire behavior models assume woody fuel particles are circular in cross section and use an assumption of a cylinder to estimate volume for other fuel properties, such as SAVR (Eq. 2.13) and density (Keane et al. 2012b). Most woody fuel particles are not cylinders, but rather, they are complicated volumes of highly variable cross sections and contorted lengths. Moreover, particle diameters are not static; they change with weather conditions, often becoming thicker when wet, and cracked when dry, making diameter measurements difficult and further complicating the estimation of SAVR. Distributions of diameters and lengths are also highly variable across woody particles. The assumptions of circular cross sections and frustum volumes are necessary due to current fire behavior modeling and fuel sampling limitations, but future efforts should explore methods for estimating SAVR and particle volume by other means.

Diameter measurements are required for many fire modeling and fuel sampling techniques (Chap. 8). Measuring particle diameter is relatively easy and is usually often done with a ruler, caliper, or diameter tape. However, many have found that these measurements are often too coarse for accurate fuel particle volume estimation, especially for fine woody fuels, because of the large variation of diameters across a fuel particle and the assumption that the particle is a cylinder or frustum (Brown 1970a). Using a single particle diameter often complicates efforts to evaluate loading sampling method accuracy and precision because a major source of uncontrolled error comes from the circular cross section assumption (Keane and Gray 2013; Sikkink and Keane 2008).

2.3.1.2 SAVR

SAVR (m^{-1}) is defined as the area of a particle surface (m^{-2}) divided by the volume of that particle (m^{-3}), but it is often indirectly estimated from particle diameter (*d*) using Eq. 2.13. Particles that are thick, such as logs, have low SAVR values (less than 1.0 m^{-1} for large logs), whereas thin particles that are long and wide, such as leaves, have high values (over 2000 m^{-1} for grass blades and pine needles). SAVR is a fuel property that indirectly characterizes particle geometry (shape), and this corresponds to the particle's importance in fire science. Particles with high SAVR (e.g., foliage) are more flammable and easier to ignite than low SAVR particles (e.g., logs; Pyne et al. 1996). SAVR indirectly represents the effect of fuel size on combustion processes. It also represents the rate of response of fuel particles to temperature and moisture fluctuations; particles with high SAVR values lose heat and moisture more quickly than particles with lower SAVR values (Brown 1970b).

SAVR is extremely difficult to measure accurately because most fuel particles are complex in geometry. The most common way to measure SAVR is to use the simple formula developed by Brown (1970b) where the fraction of particle perimeter divided by the average cross-sectional area taken for cross sections along the length of the particle. This technique requires an assumption of a geometric shape of the cross section, and most efforts assume a circle to represent the fuel particle volume, although many have used other shapes for needles, leaves, and grass blades. However, most fuel particle cross sections are difficult to describe with any general geometric shape, rather, they are complex amorphous forms. Another method is to estimate volume by submerging the particle in a liquid and measuring the displacement of the liquid, and measuring the surface area by assuming some geometric shape and measuring various dimensions to estimate area. A more complex technique would be to measure the rate of drying of the fuel particle and correlating that rate to surface area. The problem with all of these techniques to estimate SAVR or density is that particles are constantly changing in response to endogenous and exogenous biophysical processes. All fuel particles are in some state of decay, and the degree of decay and its distribution across a particle can affect SAVR. Moreover, fuel particles are constant changing shapes in response to fluctuations in moisture content, temperature, and relative humidity as mentioned above. These responses sometime result in the fragmentation of the particle, which then increases surface area and SAVR. This dynamic quality of fuel particles results in greater variability in the estimation of particle SAVR.

2.3.1.3 Particle Density (ρ_p)

Particle density is the dry weight of the particle per unit volume (kg m^{-3}). The term specific gravity is also used to represent particle density; specific gravity is the density of a substance relative to the density of water at a specific temperature and pressure. One needs to multiply specific gravity by 1000 to convert to density (e.g., 0.42 specific gravity is 420 kg m^{-3}).

Particle density is measured using variations of two techniques. The particle is always oven-dried and weighed to determine mass. Then there are two techniques for measuring volume. The first technique calculates volume by assuming various geometric shapes and using diameters and lengths to define shape dimensions (see

Sect. 2.3.1.1). However, there are great measurement errors when fuel particles are small because of highly variable fuel dimensions and inaccurate measurement procedures (Keane et al. 2012b). The other technique involves dipping the particle in liquid and calculating the displacement in volume or mass as mentioned above for SAVR. This technique is more accurate, but there are several problems that must be addressed to get more precise measurements. First, care must be taken to ensure the particle does not absorb the liquid, and this is done in a number of ways, including dipping the particle in wax or some other substance that prevents absorption or using a liquid that will not be readily absorbed by the particle. Displacement by weight can be estimated if the specific gravity of the liquid is known, while displacement of volume is somewhat problematic in that it is difficult to accurately estimate displaced volume for small and large particles.

Particle density is another property that is difficult to measure because of its high variability within a particle, across fuel types, and among fuelbeds. The density of some particles, especially woody fuels, can vary substantially along the length of the particle. Logs, for example, can be in various stages of decay along their lengths because of their contact with the ground resulting in a wide variety of densities within one particle. Most material in the litter fuel component often exists as foliar material in various states of decomposition because of their position in the vertical litter profile. And, similar to SAVR, particles are constantly changing in volume in response to environmental conditions resulting in changes in density. And each fuelbed results from a unique combination of disturbance history, vegetation development, and moisture regime, all influencing particle densities.

2.3.1.4 FMC and Moisture of Extinction (M_x)

FMC is one of the most important and dynamic fuel properties, so it is discussed in a separate chapter (Chap. 5) and will not be detailed here. Some refer to the moisture level in wildland fuel as the *fuel state* or *condition* (DeBano et al. 1998), and some refer to those fuels that can burn because they are dry enough as *available fuels* (Brown and Davis 1973). Fuel moisture provides the important link to estimate fire danger (Deeming et al. 1977), and is perhaps one of the most critical inputs in fire behavior prediction models (Andrews 1986; Table 2.2). Fuel moisture also is important to many other ecological processes, such as decomposition, evapotranspiration, and nutrient cycling.

The moisture of extinction (M_x) is the moisture content at which combustion cannot be sustained (moisture above which fire does not burn; Rothermel 1972), and greatly depends on the type, quantity, and arrangement of fuels and their interaction with weather, mainly wind. Dead woody fuels are often assigned M_x of 30 % while M_xs for live fuels are much harder to quantify. This property is actually a static parameter used in fire behavior modeling algorithms at the fuel component level (see Table 2.2) to drive combustion to zero at high moisture contents (Rothermel 1972). It would be difficult to estimate M_x under field conditions because it would change with ambient weather (e.g., temperature, humidity, incident radiation) and particle qualities (e.g., rot, density, shape, size), and live fuel plant condition (e.g., phenology, moisture stress, size). In reality, the distribution of M_x in a typical fuelbed could be quite variable and difficult to accurately quantify for operational fire management.

2.3.1.5 Mineral Content (S_e, S_T)

Mineral content (S_e and S_T depending on the equation in Table 2.2) has a profound impact on fire behavior (Philpot 1970; Eqs. 2.9, 2.10). Biomass with high mineral content, such as duff and slash fuel, will tend to burn slower and have a higher proportion of burning in smoldering combustion, often resulting in reduced combustion. In fact, fire retardant depends on this relationship to be effective; the primary purpose of retardant is to retard the spread of fire by increasing mineral content, thereby depressing fire spread (Giménez et al. 2004). The mineral content (S_T) is the percent of the total weight per unit volume of fuel particle that is inorganic material or mineral (i.e., not composed of molecules of C, H, and O). It is usually estimated by burning a fuel particle of known dry weight and weighing the ash that is left after complete combustion; the weight of ash divided by dry weight of the wood is the mineral content. The effective mineral content (S_e) is the mineral content with the proportion of silica removed (Table 2.2).

Since minerals are key nutrients needed in plant photosynthesis and respiration, they become incorporated into biomass, and each fuel particle has its own relatively static mineral content (around 5 %). Wood in woody fuels is usually 1 % mineral, while the bark can have ten times that amount (Ragland et al. 1991). Needles have fewer minerals (0.1 %), but mineral content often increases with needle age (Weikert et al. 1989). Particles near or in contact with the ground will usually have higher mineral contents because of the diverse processes involved in decomposition (Chap. 6); soil macrofauna break down organic material and often incorporate mineral soil onto the downed fuel particles. And, minerals will also tend to accumulate in the duff layer as microbes process the organic material and leave the minerals to collect in the duff (Chap. 6). As a result, ground fuels usually have the highest mineral contents, often greater than 10 %, partially explaining why ground fuels mostly burn in smoldering combustion. Keane et al. (2012b) found mineral contents were the highest in the smallest fuel particles with litter and duff having 10–50 % mineral contents, 1 h woody having 2–5 %, 1,000 h woody having 0.1–0.8 % for forest and rangelands of the northern Rocky Mountains, USA. However, these mineral contents varied greatly from site to site, and stand by history.

2.3.1.6 Heat Content (h)

The heat content (h) is the heat yield of the fuel per unit mass (kJ kg^{-1}) and, when multiplied by loading (W), is used to compute fire intensity (Eq. 2.4). This is best described as the heat released from the combustion of the gases evolved in the ignition phase (see Sect. 2.1.2) and is also called the effective heat content (Shafizadeh et al. 1977). The heat content of wildland fuel is mostly dependent on the chemical composition of the material being burned. While the majority of fuel is cellulose and lignin, there can be other chemical constituents of fuel that affect heat content. Biomass with high mineral contents, for example, will have lower heat contents (Susott et al. 1975). However, there are many chemical compounds in biomass that may increase heat content. Oils, resins, and proteins may increase heat contents in foliage and other parts of the plant (Philpot 1969). Moisture content also governs the amount of heat given off from burning fuels (Chap. 5) because heat must be used to vaporize the free and bound water in live and dead fuel particles.

Wildland fuel heat content values are quite different within and among fuel types, season, and the intensity of the fire when it is burned. Foliage usually have higher heat contents (20–21 MJ kg^{-1} or 8700–9400 BTU lb^{-1}) than twigs and stems (18–20 MJ kg^{-1} 8300–8700 BTU lb^{-1}), but this relationship is quite different across species, age, and dead versus live fuels (Philpot 1969). Kelsey et al. (1979) found that the heat content of wood ranged from 19.3 to 22.5 MJ kg^{-1} (8300–9700 BTU lb^{-1}), while bark heat content values were substantially higher ranging from 20.2 to 25.3 MJ kg^{-1} (8700–10,900 BTU lb^{-1}), and foliage heat contents were in between 20.1 and 22.4 MJ kg^{-1} (8700–9700 BTU lb^{-1}). And, the heat content might change over the course of a fire season. Philpot (1969) found that the heat content for chamise shrub leaves were lowest in the spring (~21 MJ kg^{-1} or 9100 BTU lb^{-1}) and increased to 23.5 MJ kg^{-1} (10,100 BTU lb^{-1}) in the autumn. And last, the heat content of fuels burned under flaming combustion might be quite different than when fuels are burned under smoldering combustion (Susott et al. 1975). Yet despite this high variability, most fire models use a constant value for heat content. As an example, a constant value of 18 MJ kg^{-1} (8000 BTU lb^{-1}) has been assigned to all but two of the Scott and Burgan (2005) fire behavior fuel models.

Heat content is usually measured using a bomb calorimeter using a method where a standardized measure of fuel is placed into a constant volume calorimeter and electrical energy is used to ignite the fuel. As the fuel is burning, it heats the surrounding air, which expands and escapes through a tube that heats water outside the tube. The change in the temperature of the water allows for calculating the amount of heat generated from the fuel.

2.3.1.7 Other Important Particle Properties

There are several other important fuel particle properties that are not directly used in fire behavior modeling, but they are still important in fuel science and management. Particle *shape* or the general geometry of a fuel particle is important because it is used to define a geometric form for which an equation can be used to calculate volume that is then used to estimate density and mass. Shape is also used to classify particles into fuel components and to parameterize fuel components for modeling. Particle shape is also important in fuel moisture dynamics, ignition processes, and combustion.

Particle *thermal conductivity* is a physical measure of the heat conduction potential of fuel or how fast heat can travel through fuel. Thermal conductivity has the complex units of W m^{-1} °K^{-1} or Joule sec^{-1} m^{-1} °K^{-1} or kg m sec^{-3} °K^{-1}, therefore having representations of energy, mass, time, length, and temperature. Thermal conductivity is most often used to describe solid fuel particles, but it can also be modified to describe heat transfer through porous fuelbeds, layers, or components. Estimates of thermal conductivity for duff, for example, are often used to simulate heat and temperature dynamics in the soil layer as a result of surface fire (Campbell et al. 1995). Bark thermal conductivity is used to estimate how hot and fast heat penetrates live tissue to estimate plant mortality (Reinhardt and Dickinson 2010). In general, most fire applications that require an estimate of thermal conductivity are for research purposes or specialized fire effects models (Reinhardt et al. 1997).

The *chemical content* of fuel particles is also important to most of the fuel properties presented here and also for other fuel properties that are input to some fire and fuel management applications. Oils and resins in some fuel particles may increase heat content (see Sect. 2.3.1.6; Philpot 1970), while high concentration of minerals in leaves and some wood may reduce flammability and dampen combustion (see Sect. 2.3.1.5; Whelan 1995). Other aspects of chemical composition may be important from a human health standpoint. Fuel particles might contain mercury or radioactive elements that, when burned, could create hazardous smoke emissions that might impact air quality (Canham and Loucks 1984). The relative concentrations of organic compounds, such as cellulose and lignin, influences those fuel properties that control live and dead fuel moisture dynamics, such as permeability and hygroscopy (affinity of cell walls to hold water; Chap. 5) and dictate rates of decomposition (Chap. 6).

2.3.2 *Fuel Component*

Most fuel component properties are quantified from a statistical summary of the fuel particle properties, which is often an average across a fuel component. For example, the 10 h woody fuel component (Chap. 3) is defined as downed deadwood particles with diameters greater than 0.6 cm (0.25 in) and less than 2.5 cm (1 in), so the average diameter (d) of the 10 h class is estimated from field measurements (Brown 1970a) and SAVR is estimated from d and ρ_p (Eq. 2.13). However, there are two fuel component properties that are measured directly and not estimated from particle properties.

2.3.2.1 Loading (W)

Loading is quantified as the dry weight mass of the fuelbed or fuel component per unit area. Loading estimates are reported in dry weight to eliminate moisture contributions to weight estimates, which can vary wildly over a fire season. The units used to represent loadings are quite important in fuel management because they are the context in which many people visualize the weight of fuel loads. Traditionally, loadings were assigned imperial units of tons acre^{-1}, but it is difficult for many fire professionals to envision what a ton of any fuel component looks like, let alone envision how it is distributed across an area as large as an acre. Moreover, the fuel components comprising a fuelbed have different scales of distribution (Chap. 6), so large areas are rarely needed to visually estimate the loading of fine fuels. Many studies now use SI units of kg m^{-2} for most fuel components (Keane and Dickinson 2007) because they are more easily visualized (it's easier to imagine a kilogram or 2.2 pounds of fuel over a square meter of ground than a ton of fuel over an acre). However, these units may be inappropriate for CWD and canopy fuels. Surface and canopy fuel loading measurement techniques are discussed in detail in Chap. 8.

Fuel loading is the primary fuel property discussed in this book because it is used extensively in wildland fire management for many purposes. The calculation of fire intensity, for example, demands an estimate of fuel loading (Eq. 2.4), and fire intensity is perhaps one of the most important fire behavior characteristics for estimating fire effects (Reinhardt et al. 2001). Loading is also used to estimate smoke emissions which directly affect human health and wellness. Loading often correlates to both vertical and horizontal fuel connectivity; undisturbed fuelbeds with high loadings are more likely to have greater canopy fuels and are more likely to be connected to fuelbeds with high loadings. Loading is also important for issues outside of fire science, such as habitat for small mammals, site productivity, carbon dynamics, and soil erosion.

2.3.2.2 Bulk Density (ρ_b)

The bulk density of a fuel component is the mass of the fuel component material divided by the volume of space within which it resides (Fig. 2.4a). Bulk density is different from wood or particle density (specific gravity) in that the volume includes the empty space between fuel component particles. Bulk density is often used to represent fuel arrangement in vertical dimensions; canopy profiles, for example, display the vertical distribution of canopy bulk density for crown fire modeling (Bebi et al. 2003). Past studies often used bulk density to represent *fuel porosity* (Countryman 1969).

Bulk density has a number of uses in fire management. First, it is an input to some important fire modeling programs (see Table 2.2); canopy bulk density is used in FARSITE (Finney 1998) to simulate crown fire propagation (Chap. 4) and fuelbed bulk density (ρ_b) is used to simulate surface fire intensity (Eqs. 2.9, 2.13). Bulk density can also be used to describe the rate at which heat can travel through a surface fuel layer. Another common application is in calculating loading for those fuel components that are difficult to sample. Duff, litter, shrub, herb, and tree regeneration fuel component loadings, for example, are difficult to measure operationally, so many fire specialists use the volume method to approximate loading. In this method, the depth of a fuel component is visually estimated or measured as an integrated average across an area, and multiplying this depth by the area of consideration gives the volume which the component occupies. The loading of that component can then be estimated by multiplying volume by bulk density (details are given in Chap. 8). The problem with calculating loading this way is deciding the scale at which to measure loading. Should volume be calculated for the entire

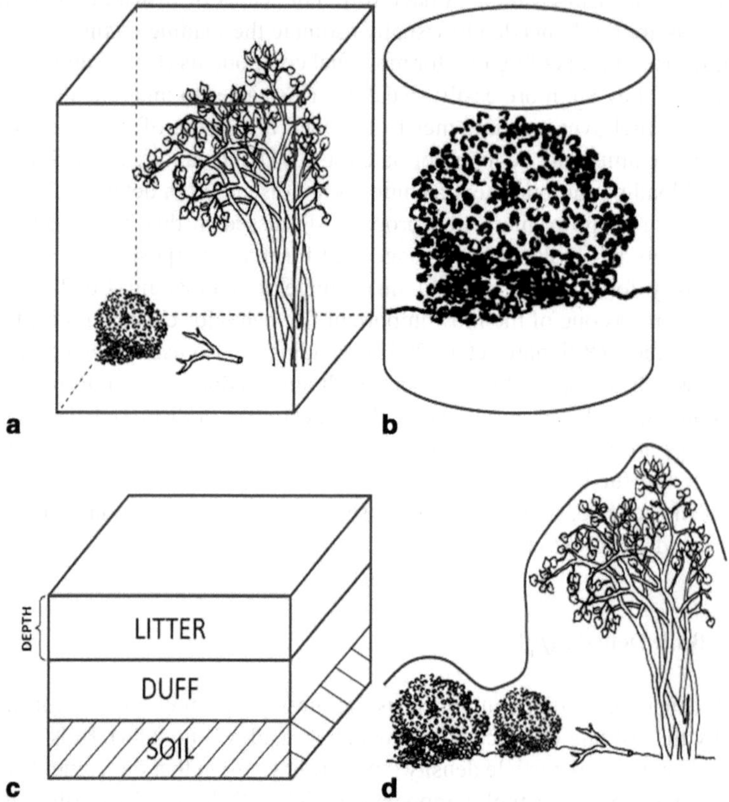

Fig. 2.4 Examples illustrating the various ways to calculate the bulk density of a fuelbed or fuel component. Bulk density is estimated as the mass of the fuel divided by volume. Volume is calculated as the area of concern times the height of the fuel making it scale dependent: **a** the volume is calculated from an estimate of depth across the fuelbed unit area; **b** the volume is calculated for the individual plant, particle, or component; **c** litter and duff bulk densities are more consistent because the depth is less variable across space; and **d** volume is calculated as an integrated average fuelbed depth

fuelbed (Fig. 2.4a), for each individual plant (Fig. 2.4b), or estimated from an average integrated height of plants in a fuelbed (Fig. 2.4d).

2.3.3 *Fuel Layer*

2.3.3.1 Fuel Layer Depth (δ)

Fuelbed depth is the thickness of the surface fuel layer (Tables 2.1 and 2.2). Many surface fuel components exist as layers of biomass above the ground, and fuelbed depth is the highest height of any fuel particle of any component integrated over the area of consideration. Fuelbed depth is an important parameter in fire behavior systems that use the Rothermel (1972) model (Andrews 1986; Andrews 2014) (Table 2.2), and, because of this, it is a parameter that is commonly adjusted to match observed with simulated fire behaviors in creating fire behavior fuel models (Chap. 7; Burgan 1987). Burgan (1987), for example, mentions that a fire behavior fuel model can be made more sensitive to wind by increasing fuelbed depth. Fuelbed depth is often used to describe only the depth of the surface fuel layer and it is mostly used to derive fuel bulk densities in US fire models (Eq. 2.10).

Fuelbed depth has little ecological value since it is so highly variable across space and time scales. Its greatest use is as input into point-level fire behavior models that simulate fire in one dimension, such as BEHAVE (Andrews 2014). Because of its scale problems and high variability, it is often difficult to obtain an accurate measurement of fuelbed depth. Initial attempts to accurately measure depth were to envision a virtual sheet over the top of the surface fuel layer and visually estimating the average height of that sheet (Jensen et al. 1993). Moreover, it is difficult to evaluate if widely spaced and distinctive fuel particles constitute part of the fuelbed. For example, should widely scattered shrubs or occasional large logs be included in the depth estimation (Fig. 2.4d). Fruiting stalks on grass and forbs, as another example, are widely scattered and are easily blown by wind because they are often taller than the plant's foliage, making it quite difficult to determine if fruiting stalks contribute to fire spread and are therefore used to estimate depth.

2.3.3.2 Packing Ratio (β)

The packing ratio is an index used to represent the compactness of the fuelbed (Rothermel 1972). It is easily quantified as the ratio ρ_b:ρ_p (fuelbed bulk density divided by particle density). This variable was invented to simulate the important damping effect of fuelbed looseness or compression on combustion using an index that is the fraction of fuelbed volume occupied by fuel. In fact, Catchpole et al. (1998) found that rate of spread decreased with the square root of the packing ratio. It had been observed that fire intensity and rate of spread occur at two extremes of compactness. Lack of fuel contagion causes loss of heat transfer in loose fuelbeds, while low air-to-fuel ratios and poor heat penetration result in lower spread rates and intensities in dense fuelbeds (Rothermel 1972). Between these two extremes is an optimum range of fuelbed packing where there is the best balance of air, fuel, and heat transfer, and this optimum packing ratio (β_{op} in Table 2.2) is greatly dependent on the fuel particles and how they are arranged in the fuelbed. Sandberg et al. (2007) modified Rothermel's (1972) equations to include a damping coefficient representing fuel compactness based on a new variable called relative packing ratio, which is the fuelbed packing ratio divided by the optimum packing ratio.

The main problem with packing ratio is that fuelbeds are often composed of many kinds of fuel particles from grass blades to woody twigs and logs; therefore, particle densities can be highly variable at very fine scales, resulting in highly variable packing ratios. Moreover, optimum packing ratios can vary across the year because of changes in plant phenology, the rate of decay, and abiotic interactions, such as grazing, trampling, and deposition (Chap. 6). As a result, the packing ratio, similar to fuelbed depth, is probably most applicable to one-dimensional fire behavior modeling and has little value in wildland fuel ecology. It would be difficult to quantify the packing ratio and its optimum in the field because of the mentioned scale issues and their temporally dynamic quality.

References

Agee JK, Skinner CN (2005) Basic principles of forest fuel reduction treatments. Forest Ecol Manage 211:83–96
Albini FA (1976) Estimating wildfire behavior and effects. USDA Forest Service, Intermountain Research Station, General Technical Report INT-30. Ogden, UT, 23 pp
Alexander ME (2014) Part 4: the science and art of wildland fire behaviour prediction. In: Scott AC, Bowman DMJS, Bond WJ, Pyne SJ, Alexander ME (eds) Fire on earth: an introduction. Wiley-Blackwell, Chichester
Anderson HE (1969) Heat transfer and fire spread. U.S. Department of Agriculture, Forest Service, Intermountain Forest and Range Experiment Station, Research Paper INT-69. Ogden, UT., USA, 20 pp
Andrews PL (1986) BEHAVE: fire behavior prediction and fuel modeling system—BURN subsystem. USDA Forest Service, Intermountain Research Station, Research Paper INT-194. Ogden UT, USA, 130 pp
Andrews PL (2014) Current status and future needs of the behave plus fire modeling system. Int J Wildland Fire 23(1):21–33
Andrews PL, Rothermel RC (1982) Charts for interpreting wildland fire behavior characteristics. US Forest Service, Intermountain Forest and Range experiment station, General Technical Report INT-131. Ogden, UT, USA, 23 pp
Barrows JS (1951) Fire behavior in northern Rocky Mountain forests. Northern Rocky Mountain Forest & Range Experiment Station, Paper 29, 274 pp
Bebi P, Kulakowski D, Veblen TT (2003) Interactions between fire and spruce beetles in a subalpine rocky mountain forest landscape. Ecology 84(2):362–371 doi:10.1890/0012-9658(2003)084[0362:ibfasb]2.0.co;2
Brown AA, Davis K (1973) Forest fire control and use, 2nd edn. McGraw-Hill, New York
Brown JK (1970a) A method for inventorying downed woody fuel. USDA Forest Service, Intermountain Research Station, General Technical Report INT-16. Ogden, UT, 16 pp
Brown JK (1970b) Ratios of surface area to volume for common fire fuels. Forest Sci 16:101–105
Burgan RE (1987) Concepts and interpreted examples in advanced fuel modeling. USDA Forest Service, Intermountain Research Station, General Technical Report INT-238. Ogden, UT, USA, 40 pp
Byram GM (1958) Some basic thermal processes controlling the effects of fire on living vegetation. USDA Forest Service Southeastern Forest Experiment Station, Research Note 114. Asheville, NC, USA, 2 pp
Byram GM (1959) Combustion of forest fuels. In: Brown KP (ed) Forest fire: control and use. McGraw-Hill, New York, p 584
Campbell GS, Jungbauer JD, Bristow KL, Hungerford RD (1995) Soil temperature and water content beneath a surface fire. Soil Sci W 159(6):363–374
Canham CD, Loucks OL (1984) Catastrophic windthrow in the presettlement forests of Wisconsin. Ecology 65(3):803–809. doi:10.2307/1938053
Catchpole WR, Catchpole EA, Butler BW, Rothermel RC, Morris GA, Latham DJ (1998) Rate of spread of free-burning fires in woody fuels in a wind tunnel. Combust Sci Technol 131:1–37

Countryman CM (1969) Fuel evaluation for fire control and fire use. Paper presented at the Symposium on fire ecology and control and use of fire in wildland management, Tucson, AZ, 1969
Curry JR, Fons WL (1938) Rate of spread of surface fires in the Ponderosa pine type of California. J Agr Res 57(4):239–267
DeBano LF, Neary DG, Ffolliott PF (1998) Fire's effect on ecosystems. Wiley, New York
Deeming JE, Burgan RE, Cohen JD (1977) The National fire danger rating system—1978. USDA Forest Service Intermountain Forest and Range Experiment Station, General Technical Report INT-39. Ogden, Utah, USA, 63 pp
Finney MA (1998) FARSITE: Fire area simulator—model development and evaluation. United States Department of Agriculture, Forest Service Rocky Mountain Research Station, Research Paper RMRS-RP-4. Fort Collins, CO. USA, 47 pp
Finney MA, Cohen JD, McAllister SS, Jolly WM (2013) On the need for a theory of wildland fire spread. Int J Wildland Fire 22(1):25–36
Fons WL (1946) Analysis of fire spread in light forest fuels. J Agr Res 73(3):93–121
Frandsen WH, Andrews PL (1979) Fire behavior in nonuniform fuels. US Department of Agriculture, Forest Service, Intermountain Forest and Range Experiment Station, Research Paper INT-232. Ogden, UT, USA, 34 pp
Giménez A, Pastor E, Zárate L, Planas E, Arnaldos J (2004) Long-term forest fire retardants: a review of quality, effectiveness, application and environmental considerations. Int J Wildland Fire 13(1):1–15
Gisborne HT (1927) The objectives of forest fire-weather research. J Forest 25(4):452–456
Gisborne HT (1947) Fundamentals of fire behavior. Fire Control Notes 9(1):13–24
Hawley LF (1926) Theoretical considerations regarding factors which influence forest fires. J Forest 24(7):756–763
Jensen ME, Hann WJ, Keane RE, Caratti J, Bourgeron PS (1993) ECODATA—A multiresource database and analysis system for ecosystem description and evaluation. In: Jensen MEaPBE (ed) Eastside forest ecosystem health assessment volume II—ecosystem mangement: principles and applications, 1993. USDA Forest Service Pacific Northwest Research Station General Technical Report PNW-GTR-318, pp 203–217
Keane R, Gray K, Bacciu V, Leirfallom S (2012a) Spatial scaling of wildland fuels for six forest and rangeland ecosystems of the northern Rocky Mountains, USA. Landscape Ecol 27(8):1213–1234. doi:10.1007/s10980-012-9773-9
Keane RE, Finney MA (2003) The simulation of landscape fire, climate, and ecosystem dynamics. In: Veblen TT, Baker WL, Montenegro G, Swetnam TW (eds) Fire and global change in temperate ecosystems of the western Americas, vol Ecological Studies vol 160. Springer-Verlag, New York, pp 32–68
Keane RE, Dickinson LJ (2007) The Photoload sampling technique: estimating surface fuel loadings using downward looking photographs. USDA Forest Service Rocky Mountain Research Station, General Technical Report RMRS-GTR-190. Fort Collins, CO, 44 pp
Keane RE, Gray K (2013) Comparing three sampling techniques for estimating fine woody down dead biomass. Int J Wildland Fire 22(8):1093–1107
Keane RE, Drury SA, Karau EC, Hessburg PF, Reynolds KM (2010) A method for mapping fire hazard and risk across multiple scales and its application in fire management. Ecol Model 221:2–18
Keane RE, Gray K, Bacciu V (2012b) Spatial variability of wildland fuel characteristics in northern rocky mountain ecosystems. USDA Forest Service Rocky Mountain Research Station, Research Paper RMRS-RP-98 Fort Collins, Colorado USA, 58 pp
Kelsey RG, Shafizadeh F, Lowery DP (1979) Heat content of bark, twigs, and foliage of nine species of western conifers. U.S. Department of Agriculture, Forest Service, Intermountain Research Station Research Note INT-261. Ogden, UT, 7 pp
Linn RR (1997) A transport model for prediction of wildfire behavior. Ph. D. dissertation, New Mexico State University, Las Cruces, New Mexico, USA
Liodakis S, Bakirtzis D, Dimitrakopoulos A (2002) Ignition characteristics of forest species in relation to thermal analysis data. Thermochim Acta 390(1–2):83–91

Mitchell SJ (2013) Wind as a natural disturbance agent in forests: a synthesis. Forestry 86(2):147–157. doi:10.1093/forestry/cps058
Moritz MA, Morais ME, Summerell LA, Carlson JM, Doyle J (2005) Wildfires, complexity, and highly optimized tolerance. Proc Natl Acad Sci 102(50):17912–17917
Ottmar RD, Sandberg DV, Riccardi CL, Prichard SJ (2007) An overview of the fuel characteristic classification system—quantifying, classifying, and creating fuelbeds for resource planning. Can J Forest Res 37:2383–2393
Parsons RA, Mell WE, McCauley P (2010) Linking 3D spatial models of fuels and fire: effects of spatial heterogeneity on fire behavior. Ecol Model 222(3):679–691
Philpot CW (1969) Seasonal changes in heat content and ether extractive content of chamise. U.S. Department of Agriculture, Forest Service, Intermountain Research Station Research Paper INT-61. Ogden, UT, 10 pp
Philpot CW (1970) Influence of mineral content on the pyrolysis of plant materials. Forest Sci 16(4):461–471
Pyne SJ (2001) The fires this time, and next. Science 294:1005–1006
Pyne SJ, Andrews PL, Laven RD (1996) Introduction to wildland fire, 2nd edn. Wiley, New York
Ragland KW, Aerts DJ, Baker AJ (1991) Properties of wood for combustion analysis. Bioresource Technol 37(2):161–168
Reinhardt E, Dickinson M (2010) First-order fire effects models for land management: overview and issues. Fire Ecol 6(1):131–142
Reinhardt E, Keane RE, Brown JK (1997) First order fire effects model: FOFEM 4.0 user's guide. USDA Forest Service, Intermountain Research Station General Technical Report INT-GTR-344, 65 pp
Reinhardt ED, Keane RE, Brown JK (2001) Modeling fire effects. Int J Wildland Fire 10:373–380
Rothermel RC (1972) A mathematical model for predicting fire spread in wildland fuels. United States Department of Agriculture, Forest Service, Intermountain Forest and Range Experiment Station, Research Paper INT-115, Ogden, Utah, 88 pp
Rothermel RC, Deeming JE (1980) Measuring and interpreting fire behaviour for correlation with fire effects. US Forest Service Intermountain Forest and Range Experiment Station, General Technical Report 93, Ogden, UT, 8 pp
Sandberg DV, Riccardi CL, Schaaf MD (2007) Reformulation of Rothermel's wildland fire behaviour model for heterogeneous fuelbeds. Can J Forest Res 37(12):2438–2455
Scott J, Burgan RE (2005) A new set of standard fire behavior fuel models for use with Rothermel's surface fire spread model. USDA Forest Service Rocky Mountain Research Station, General Technical Report RMRS-GTR-153. Fort Collins, CO, 66 pp
Shafizadeh F, Chin PPS, DeGroot WF (1977) Effective heat content of green forest fuels. Forest Sci 23(1):81–89
Sikkink PG, Keane RE (2008) A comparison of five sampling techniques to estimate surface fuel loading in montane forests. Int J Wildland Fire 17(3):363–379. doi:10.1071/Wf07003
Sullivan AL (2009a) Wildland surface fire spread modelling, 1990–2007. 1: physical and quasi-physical models. Int J Wildland Fire 18(4):349–368
Sullivan AL (2009b) Wildland surface fire spread modelling, 1990–2007. 2: empirical and quasi-empirical models. Int J Wildland Fire 18(4):369–386
Susott RA, DeGroot WF, Shafizadeh F (1975) Heat content of natural fuels. J Fire Flammability 6:311–325
Thomas PH (1953) Effects of fuel geometry in fires. Building Research Establishment Current Paper. Department of the Environment, Building Research Establishment, Borehamwood, 15 pp
Trakhtenbrot A, Katul GG, Nathan R (2014) Mechanistic modeling of seed dispersal by wind over hilly terrain. Ecol Model 274(0):29–40
Van Wagner CE (1983) Fire behaviour in northern conifer forests and shrublands. In: Wein RW, MacLean DA (eds) The role of fire in northern circumpolar ecosystems. Wiley, Chichester, pp 65–80
Weikert RM, Wedler M, Lippert M, Schramel P, Lange OL (1989) Photosynthetic performance, chloroplast pigments, and mineral content of various needle age classes of spruce (Picea abies)

with and without the new flush: an experimental approach for analysing forest decline phenomena. Trees 3(3):161–172. doi:10.1007/bf00226652

Whelan RJ (1995) The Ecology of Fire. Cambridge studies in ecology. Cambridge University Press, Cambridge

Zhou XY, Mahalingam S (2001) Evaluation of reduced mechanism for modeling combustion of pyrolysis gas in wildland fire. Combust Sci Technol 171:39–70. doi:10.1080/00102200108907858

Chapter 3
Surface and Ground Fuels

Wood already touched by fire is not hard to set alight
African proverb

3.1 Introduction

Surface fuels always get the most attention in fire management because it is the fuel layer that is most often used to predict fire behavior. Most wildland fires spread through surface fuels so the majority of the world's fire behavior prediction packages demand detailed physical descriptions of surface fuel components for fire simulation. In fact, it is rare to have only a ground or crown fire without a fire also burning in the surface fuel. Ground fuels were included in this chapter because they are important to smoldering combustion; they often comprise the majority of fuel loading in many forest ecosystems; and they are often included with surface fuels in various sampling, classification, mapping, and simulation efforts (Lutes et al. 2009; Riccardi et al. 2007a).

3.2 Surface Fuels

The surface fuel layer contains the most diverse and complex fuel types in fire management. Surface fuel particles are often dissimilar physically across most of the common surface fuel types, especially considering that some are live and some are dead. Herbaceous particles (e.g., grass blades), for example, have significantly different properties (e.g., shape, particle density) than downed wood (e.g., branches and logs). There are a core set of seven surface fuel components that are common inputs to most fire models and they are also present in most surface fuel beds of the world, especially forests (Table 3.1, Fig. 1.2). Many ecosystems have the surface fuel types of litter, shrubs, and herbaceous fuels in varying amounts above the ground surface. Forest and shrub ecosystems nearly always have woody fuels in a wide range of abundance and sizes.

R. E. Keane, *Wildland Fuel Fundamentals and Applications*,
DOI 10.1007/978-3-319-09015-3_3

Table 3.1 Descriptions of the seven most common surface fuel components and the one most common ground fuel component found in ecosystems worldwide. FWD is fine woody debris, a term often given to wood fuel particles less than 8 cm in diameter. CWD is coarse woody debris, a term used to woody fuel particles greater than 8 cm in diameter

Fuel type	Fuel component	Common name	Size	Description
Surface fuels				
Downed dead woody	1-h woody	Twigs, FWD	<0.6 cm (0.25 in) diameter	Detached small woody fuel particles on the ground
	10-h woody	Branches, FWD	0.6–2.5 cm (0.25–1.0 in) diameter	Detached small woody fuel particles on the ground
	100-h woody	Large branches, FWD	2.5–8 cm (1–3 in) diameter	Detached small woody fuel particles on the ground
	1000-h woody	Logs, CWD	8+ cm (3+ inch) diameter	Detached small woody fuel particles on the ground
Shrubs	Shrub	Shrubby	All shrubby material less than 5 cm diameter	All burnable shrubby biomass with branch diameters less than 5 cm
Herbaceous	Herb	Herbaceous	All sizes	All live and dead grass, forb, and fern biomass
Litter	Litter	Litter	All sizes, excluding woody	Freshly fallen nonwoody material which includes leaves, cones, pollen cones
Ground fuels				
Duff	Duff	Duff	All sizes	Partially decomposed biomass whose origins cannot be determined

FWD fine woody debris, *CWD* coarse woody debris

3.2.1 Litter

Litter is considered freshly fallen, readily identifiable dead plant biomass (Fig. 3.1a). It is often called the L layer in vegetation and soils studies (Soil Classification Working Group 1998). Of all the surface fuel components, litter is often the most diverse because it may consist of a wide variety of fallen plant parts, such as bud scales, pollen cones, and dead grass blades. However, the most common constituent of litter is plant foliage, often needles and leaves, and these discarded plant parts readily facilitate the spread of fire. In ecology, litterfall is a term often used to describe the biomass dropped by plants and so it follows that the litter layer may contain all types of plant material, including woody particles. But, in fuel science, litter excludes woody material as woody fuels are a separate component with completely different physical, chemical, and combustion properties (Table 3.1). Litter particles are often small, light, and thin, such as needles, leaves, and grass blades, which mean they have high SAVRs, and the particles are often loosely distributed above the duff (high packing ratios and low bulk densities). Therefore, litter material is able to dry quickly and is easily ignited to spread fire, and it burns mostly in the flaming combustion phase (Chap. 2).

The high diversity of litter fuel particles makes the differentiation and measurement of the litter layer properties difficult. In fact, many litter particle types could be their own fuel component in more refined fire modeling applications, but these particles are often included in the litter component for ease of operational measurement and simulation modeling. Cones, for example, are often included in the litter layer for simplicity (Fig. 3.1b), yet cones smolder for long times and could impart significantly different thermal effects on the ground than many of the other litter particles. One consistent quality of litter is that fallen litter particles are often small and evenly distributed across the fuel bed. However, some fallen plant parts, such as cones, buds, and most importantly bark, are large and, if they are included in the litter, they often have a more heterogeneous spatial distribution and more disparate fuel properties (Keane 2008). To include these odd particles in the litter component for a specific fire application may add to greater uncertainty into any sampling effort or inventory protocol used to estimate loading. Yet, these particles contribute to fire spread and may affect other fire behavior attributes, such as spread rate, intensity, and residence time. Another major problem is identifying where the litter layer stops and the duff layer begins because of the gradient of decomposition from the litter to mineral soil (Fig. 3.2a) (see Sect. 3.3.1 for more detail). In the end, the set of elements and particles that define a litter fuel component must be decided based on the objective of the particular fuel application and compromises must be made to facilitate cost-effective sampling (Lutes et al. 2006; Chap. 8).

3.2.2 Woody Fuels

All forest, woodland, and shrub ecosystems have dead woody fuels in the surface fuel layer (Fig. 3.1c, d). Woody fuels are those twigs, branches, and boles of woody plants (mostly trees, shrubs, and lianas) that are dislodged and drop to the ground. These woody particles are mostly composed of cellulose and lignin. Woody fuel types are important to fire science in that they can foster intense wildland surface fires, so their study has dominated much of fire and fuels research. In an operational sense, a surface fuel particle is classified as woody fuel if it meets three criteria—particles must be down, dead, and woody—hence the name "down dead woody fuels" used throughout the wildland fuel literature. Particles are *downed* if they are detached from their parent plant and are below the 2-m surface fuel layer (Fig. 1.2, Fig. 3.1c, d). Some studies considered attached dead branches below 2 m as part of the fuel bed for certain measurement objectives, but generally, downed fuels must

Fig. 3.1 Examples of the primary surface fuel components: **a** litter above the duff ground fuel layer, **b** litter, cones, and woody (ponderosa pine forest in southern Idaho, USA), **c** 1- and 10-h-downed dead woody fuels (Douglas-fir/ponderosa pine forest in northwest Montana, USA), **d** 10- and

be unattached. Surface woody fuels must also be *dead,* not living. This is important because dead woody fuel's physical properties, such as moisture, react differently to exogenous factors, such as temperature, precipitation, and relative humidity, than live fuels (Chap. 5). It is sometimes difficult to determine whether woody fuels are live or dead, especially when live branches are partially covered by the duff and litter layers, so when there is a question, it is often necessary to break the particle in half and ensure living tissue is present to verify that it is in fact live. And last, the particles should be *woody*. This is often difficult to tell in the field because some forbs produce structural plant stalks that look quite similar in appearance to woody fuels. However, you can tell if particles are woody simply by picking them up and breaking them to ensure that there is a woody center. Nondendritic plants will often be hollow or contain a nonwoody center.

In most fire management applications, woody fuel types are divided into fuel components based on their size as determined by their diameter (Table 3.1). Historically, ranges of woody fuel particle diameters were based on their rate of drying rather than some other ecological characteristic (Fosberg et al. 1970). So, to understand conventional stratifications of woody fuels, one has to understand the concept of time lag in dead fuels. *Time lag* is the time it takes for a fuel particle to lose 63 % of the difference between its initial moisture content and its equilibrium moisture content under specific environmental conditions (80° F, 20 % relative humidity) assuming an exponential drying function (see Chap. 5; Fosberg et al. 1970). Therefore, based on time lags, woody fuel particles were stratified into four standardized size classes; the 1-h woody fuels are those fuel particles that lose approximately two thirds of their moisture in 60 min and this corresponds to particle diameters that are less than 0.6 cm (0.25 in) in diameter (Fig. 3.1c). The 10-h woody fuels are particles with diameters between 0.6 and 2.5 cm (0.25–1 in) and 100-h woody fuels are 2.5–8.0 cm (1–3 in) in particle diameter (Fig. 3.1d, e). Woody particles with diameters greater than 8 cm are called 1000-h fuels (Fig. 3.1f).

One consequence of defining woody fuel components by rate of moisture loss is that the particle diameter size classes are inappropriately designed to get accurate assessments of fuel properties and characteristics. Woody fuel properties can vary widely across the particle sizes comprising a woody fuel component. Fuel moistures can also be highly variable within a size class; smaller particles within a size class can be much drier on any given day. This also matters in the calculation of loading. The 100-h time lag woody fuel component, for example, has large differences in fuel biomass between a 3.5-cm diameter branch and an 8-cm branch because particle mass is often computed as the product of particle volume and density, and since volume increases with the square of the diameter, small particles in the 100-h class can have nearly nine times more volume than the larger particles of the same length (2.5–8.0 cm). This makes the accurate determination of loading difficult because

100-h fuel (lodgepole pine forest in central Montana, USA), **e** 100-h woody fuel at 5.0 kg m^{-2} loading (Keane and Dickenson 2007), **f** logs or 1000-h fuel (lodgepole pine forest central Idaho, USA), **g** shrub (snowberry in southwestern Montana, USA), and **h** herbaceous (eastern Washington USA grassland, photo courtesy of US Forest Service FERA)

the large diameter class ranges causes high amounts of uncertainty in sampling estimates (Keane and Gray 2013). Moreover, biophysical processes that control fuels, such as decomposition and deposition (Chap. 6), vary greatly across the particle sizes within the conventional 10-, 100-, and 1000-h woody fuel size classes. Therefore, to improve future fuel assessments, a new size classification is needed that stratifies woody fuels according to ecological characteristics rather than their drying rate.

Woody fuels are also often divided into two general fuel types. *Fine woody debris* (FWD) are generally those woody fuel particles less than 8 cm (3 in) in diameter, while *coarse woody debris* (CWD) are fuel particles greater than or equal to 8 cm diameter (Fig. 3.1c, d, e). FWD includes the 1-, 10-, and 100-h woody fuel components which are the input fuel components in the US fire behavior models. FWD is important in fire dynamics because they both facilitate fire spread and contribute to fire intensity.

Coarse woody debris (CWD) consists of fallen logs on the forest floor that fire managers and scientists refer to as 1000-h woody fuels (8 + cm diameter; Table 3.1, Fig. 3.1f). CWD also has size classes but they are not standardized across all fire management applications; diameter ranges for CWD fuel components are designed for specific purposes (Riccardi et al. 2007a). Some fuel classifications divide the CWD into three fuel components: particle size diameter classes of 8–23 cm (3–9 in), 23–51 cm (9–20 in), and 51 + cm (20 in) that define 1000, 10,000, and 10,000+ h time lag fuels respectively (Riccardi et al. 2007a). As far as fire behavior is concerned, 1000-h fuels contribute little to fire spread (Anderson 1969), so finer divisions of CWD size classes were unnecessary for the simulation of fire propagation so all CWD biomass was grouped into one class regardless of size, length, or condition of the fuel particle. However, the CWD fuel component may often have the highest loadings in forested settings (Brown and Bevins 1986; Table 3.1, Fig. 3.2b), so CWD can contribute to high fire intensities. Moreover, larger logs burn or smolder longer than smaller logs that usually results in more heat pulsed into the ground (Albini and Reinhardt 1995). If fuel inventories and data are to be useful to other resource applications, such as wildlife habitat determination, fuel consumption, and smoke emissions modeling, there must be finer divisions of CWD size classes (Albini and Reinhardt 1995).

An important quality of woody fuel is the degree of rot (decomposition) in the woody fuel particle. The level of rot in woody fuels affects many other fuel properties. As woody fuels rot, for example, there are increases in SAVR and decreases in particle density and heat content. These changes also affect drying and wetting rates, ignitability, and fire intensity (Pyne et al. 1996). Rotten wood, for example, often burns in smoldering combustion and can ignite at higher moistures than sound wood. As a result, woody fuel types and components are also described by their degree of rot (Chap. 8).

Fig. 3.2 Examples of some important fuel complexes and their primary fuel components: **a** litter (ponderosa pine stand in Lassen Volcanic National Park, California, USA, photo courtesy of Sharon Hood), **b** woody (lodgepole pine forest Yellowstone National Park, photo courtesy of USFS FERA), **c** shrub (northern US California chaparral, photo courtesy of USFS FERA), **d** shrub (pocosin shrubland North Carolina, USA, photo courtesy of Jim Reardon), **e** herbaceous (tussock grasslands of New Zealand), and **f** herbaceous (*Sorghum* spp. grass layer in northern Australia tropical savanna)

3.2.3 Shrubs

The shrub fuel type consists of nontree, woody, and foliage vegetation biomass (Fig. 3.1g). A shrub fuel type can dominate a fuel bed where the loading of shrub biomass overwhelms the loading of all other components to become a shrub fuel complex (Fig. 3.2c), or shrubs can be a minor component in the surface fuel layer (Fig. 3.1g). Shrub fuel types are often stratified by live and dead components, but most fire applications consider shrubs mostly live fuels, even though there are many shrub species of the world that retain their dead leaves after heavy frosts (e.g., turkey oak). Like litter, shrubby fuels are also diverse in terms of size, type, and distribution of fuel particles, but most fire applications put all shrub material (foliage, twigs, branches) into only one component. The woody material in shrub stems is rarely stratified into components based on diameter size classes, and foliage particles are rarely separated into finer species, size, or shape components. Some studies have suggested dividing shrub fuels into two components based on size (above and below 5 cm (2 in) in diameter, Brown et al. 1982), but few fire models accept these components. Frandsen (1983) divided live and dead sagebrush fuels into size classes that correspond to fire behavior inputs: 0–0.6, 0.6–2.0, and 2.0–7.0 cm diameter.

Some of the most important fuel beds in the world are shrub fuel complexes (Fig. 3.2c). Shrub-dominated communities are often stratified into several structural types: *scrub* (dense foliar cover >30%, 2–8 m tall), *tall shrub* (sparse foliar cover <30%, 2–8 m tall), *heath* (dense foliage with cover >30%, 0–2 m tall), and *low shrub* (sparse foliage with cover <30%, 0–2 m tall). Sclerophyllous shrublands, such as the chaparral ecosystems of western North America, the bushlands of Australia, the garrigue and maquis communities of the Mediterranean, and the fynbos of southern Africa, are some of the most interesting shrub ecosystems because they have several characteristics in common. First, they all have a Mediterranean climate that has a pronounced summer dry season and winter rains. The vegetation is dominated by shrubs with hard leaves that have short internodes (sclerophyll) and these ecosystems also are diverse in species composition. Soils are often poor and productivity is commonly low in most of these communities. But most importantly, these sclerophyll shrublands have a history of frequent burning, either by aboriginal populations or by lightning, and most of the shrub species has evolved diverse adaptations for surviving fire.

3.2.4 Herbaceous

Herbaceous fuel types are probably the most interesting of all the fuels because they consist of biomass from nonwoody plants such as grasses, forbs, and ferns, and not trees, shrubs, mosses, and lichens. This diverse group includes many species of grasses, sedges, rushes, forbs, and ferns; an eclectic group of life-forms and species that span a wide variety of ecological characteristics and combustion properties (Fig. 3.1h). Grasses, sedges, and rushes, for example, have leaves that are long and linear with high surface area to volume ratios that create loosely packed fuel beds making them highly flammable. On the other hand, forbs and ferns can have foliage (large leaves) and structural tissue (stems) that might make them more difficult to ignite. The species in this diverse fuel type are vastly different in terms of phenology, morphology, and life cycle stages resulting in a wide range of fuel conditions within a fuel bed on any day of the year. Some species cure before or during the fire season to become highly flammable, while some stay alive the entire year without the major changes in flammability. Some annuals don't appear till later in the year, while others appear only in the spring. And, since most aboveground herbaceous biomass eventually dies over the course of a year, the populations and amount of biomass produced by these species is heavily governed by climate; wet years produce greater amounts of biomass that then can be burned after curing or during the coming dry year. The study of the herbaceous fuel beds is difficult because the large number of species that comprise the fuel component creates a gradient of fuel condition (loading, moisture) that varies each day and across each year.

The herbaceous fuel complex is best known for its high flammability in that fires in herbaceous fuels spread quickly, especially in high winds (Fig. 3.2f and e). Grass fuels were responsible for the fast-spreading fire that killed 13 men in Mann Gulch in central Montana USA in 1949 (Rothermel 1993). There are several reasons why herbaceous fuels are so flammable. First, they often cure as the fire season advances. *Curing* is a measure of "greenness" and the degree of curing is defined here as the percentage of the aboveground herbaceous material that is dead (Flavelle 2008). Second, herbaceous fuels are morphologically suited for easy ignition and fast spread for wildland fires; most fuel particles are small and thin, with high SAVRs. The leaves of many herbs, especially grasses, stay erect after curing to form loose fuel beds that conduct surface fire easily. Many grasslands and herblands of the world are loosely packed contiguous surface layers that are at bulk densities that foster rapid spread of fire. Herbaceous fires are mostly of high-intensity burning mostly in the flaming stage but of short duration, but there are some grasslands that have a buildup of thatch (undecomposed herbaceous litter) that may burn in smoldering combustion.

Many areas of the world are dominated by grass and herbaceous fuel complexes (Fig. 3.1h). The Great Plains of North America, the outback of Australia, and the pampas of South America are all examples of grasslands with a history of frequent fire and where fire management is important. Savannas also contain herbaceous fuel complexes and are perhaps the most interesting in that the grassland is dotted with widely scattered trees that are able to remain on the landscape because they are adapted to both survive fire and regenerate after fire (Higgins et al. 2000; Fig. 3.2f).

3.2.5 Special Surface Fuel Types

There are some fuel types that are either too sparsely distributed to be considered a separate fuel component or unique to a particular ecosystem or disturbance regime and therefore are not considered in most fire models. These special fuel types warrant attention because they represent a significant fuel source in some areas and they can't be grouped into the seven common surface fuel types mentioned above. Historically, fuel inventories have ignored them, lumped them into the common fuel types, or measured them separately using specialized sampling techniques. Again, the recognition of special fuel types for local evaluations would ultimately depend on the sampling objective. If smoke emissions were a concern, for example, then it would be important to get loading estimates of everything that can burn, including the special fuel types, but if fire spread predictions were desired, then the quantification of special fuel types that don't contribute to fire spread becomes less important. The following are some examples of special fuel types, stratified by their origins, that may be important in some areas. This list is by no way exhaustive since there are many more special fuel types in the ecosystems of the world.

3.2.5.1 Plant Origin

Most fuel bed descriptions do not include *snags* in CWD fuel components, although there are some studies that consider snags as CWD (Harmon et al. 1986). A snag is a standing dead tree, but sometimes it is difficult to determine if a snag is a log, especially when it is leaning (Fig. 3.3a). When trees or snags fall, they often get caught or hung up against other trees such that the fallen log looks very much like an upright snag. This poses a dilemma because snags contribute little to fire spread or intensity, but logs are important in the calculation of fire intensity. From the fuel perspective, snags are tomorrow's 1000-h fuel, but they contribute little to fire intensity and spread when they are erect. In a biological sense, snags are completely different from logs and have their own inherent ecological value, such as providing nesting and foraging habitat for a number of common to rare avian wildlife, including the spotted owl and pileated woodpecker (Hutto 2006). Snags also contain unique assemblages of other organisms (Nappi et al. 2004). Therefore, to lump snags in with CWD seems inappropriate for both ecological and physical reasons. So, when does a snag behave like a log when it is burned? In this book, CWD are considered downed if the long axis of the log is at an angle greater than 45° from the vertical (<45° from ground line). This distinction is more related to combustion physics than ecology or any other field of study because the ecosystem processes acting on a snag regardless of its angle are completely different from those acting on a log that is in contact with the ground. Moreover, snags are often more dangerous to firefighters than logs because snags may fall, especially when fire burns around the trunk, and the falling material can hurt people. A quantification of the snag component might be important, not so much for fire modeling, but as an additional assessment of fire risk and hazard, and for ecological concerns.

Mosses and lichens are present in many ecosystems but at such low amounts that they often contribute little to fire spread and intensity. However, there are extensive land areas of the world where live and dead lichen and moss can accumulate to create thick litter layers that foster intense fires (Fig. 3.3b). Mosses and lichens, whether they are dead or alive, have dramatically different moisture dynamics and

Fig. 3.3 Examples of some special fuel types: **a** whitebark pine snags (central Idaho, USA, photo courtesy of Steve Arno), **b** moss and lichen (Canadian boreal forest, photo courtesy of Dan Thompson), **c** Douglas-fir tree regeneration (Missoula, MT, USA), **d** squirrel midden (cache of whitebark pine cones in Yellowstone National Park, USA), **e** slash (a grand fir selective harvest in north Idaho, USA), **f** masticated (treated stand in northern Idaho, USA), and **g** organic fuels (peat bog, Agassi Wildlife Refuge, Minnesota USA; photo courtesy of Jim Reardon)

fuel properties than wood, shrub, or herbaceous fuels (Sylvester and Wein 1981; Chap. 5). In general, mosses and lichens have high SAVR, low heat content, and low particle density that allow for their ability to dry quickly and facilitate fire spread. Many fires in the boreal forests of North America and Europe are spread through surface fuels that are primarily live and dead feather and sphagnum mosses (Shetler et al. 2008).

One fuel type or component that often gets ignored in many fuel and fire applications is tree regeneration (tree seedlings below 2 m from the ground; Fig. 3.3c). Tree seedlings are sometimes inappropriately included in the shrub and herbaceous fuel components for simplicity, but tree regeneration differs from shrubs in leaf morphology, particle types and sizes, and moisture dynamics. Sometimes, tree seedlings are included in the canopy fuels (Chap. 4), but most of this biomass occurs within the surface fuel layer and will contribute to surface fire spread under the right conditions. That said, tree seedling loadings are not inputs into most fire behavior and effects models. Future fire models should endeavor to include tree regeneration as a unique fuel component and future sampling efforts should include a quantification of tree seedling loadings in their design.

Dead plant parts that fall and get hung up on living and dead plants are also difficult to categorize. When foliage and branches fall, they sometimes catch on upright plants and remain there throughout the fire season or maybe longer. "Needle drape" is well known in the southwestern USA where fallen dead needles from productive pine stands get caught in understory shrub and understory tree layers to create a surface fuel layer that is highly flammable (Andreu et al. 2012).

Litter accumulation around the base of trees is another specialized case of litter fuels. Needles and bark often accumulate around large fire-adapted trees in stands that haven't experienced fire in a long time (Fig. 3.1b). These litter mounds can cause substantial damage to the central tree when they eventually burn either by prescribed burns or by wildfire. Few litter sampling protocols include modifications in sampling methods or intensities to sufficiently estimate the loadings and properties of these litter mounds.

3.2.5.2 Animal Origin

An important fuel type that is present in many sylvan communities is squirrel middens (Fig. 3.3d). Middens are places where squirrels have cached cones over generations. These middens consist of intact cones, cone scales, and cone cores. Some middens can be quite large (2–5 m in diameter) but they are widely scattered across the landscape. These piles burn more like duff than litter in that they mostly smolder during a surface fire. Middens may contain up to 5–10 kg m^{-2} of organic material that could be an important carbon pool (Riccardi et al. 2007b).

Animal scat or droppings is another interesting fuel type that can be quite diverse. Scat, when burned, also smolder, pulsing heat deeper into the soil profile and sometimes causing high mortality in soil biota. Again, most fuel applications lump wildlife droppings in with the litter, but areas with high scat densities may need to have a more comprehensive description and quantification of these unique fuel types.

3.2.5.3 Disturbance Origin

Many distinctive fuel beds were created by exogenous disturbances. Exogenous disturbances are disturbance events that originate from outside of the stand or fuel bed, such as fires, hurricanes, and avalanches, and spread through the area to alter fuel bed characteristics (see Chap. 6). High wind events can topple trees over large areas and create unique blowdown of fuel beds where logs are aligned along wind direction (Woodall and Nagel 2007). Hurricanes can blow trees and branches down to the ground and create fuel beds that may foster other disturbances, such as fire and insect outbreaks (Busing et al. 2009; Fig. 1.2b). While many insects and diseases may damage or kill occasional plants when at low population levels (endemic level), outbreaks of these insects and pathogens can occur under the right climatic and vegetation conditions and these outbreaks can severely damage or kill plants to create fuel beds that many consider hazardous (Jenkins et al. 2008). Mountain pine beetles, for example, bore into the cambium of host pines and lay their eggs and also introduce fungus in tree tissue. Both the fungus and the larvae that hatch from the eggs damage or kill the tree, and if the beetle population is at epidemic levels, many trees will be killed to alter both canopy and surface fuels, from both falling dead biomass and the facilitation of canopy growth of surviving competitors (Fig. 1.2c). It would be impossible to describe the fuel beds created by all possible exogenous disturbances since it would greatly depend on pre-disturbance stand conditions, antecedent climate, disturbance intensity, and a host of other biophysical factors. However, it is critical to acknowledge the importance of severe exogenous disturbances in influencing fuel dynamics (Chap. 6).

3.2.5.4 Human Origin

Several important fuel complexes result from land management activities. First, the slash fuel bed is created when the vegetation is treated using a wide variety of techniques depending on land management objectives (Fig. 3.3e). Slash is the biomass left on a treated site. It is often branches, needles, and other material that isn't merchantable or of little valuable to society. The amount of slash left after harvesting or silvicultural activities greatly depends on the density of the treated vegetation, the specifics of the slashing treatment, the amount of material removed from the site, and the equipment used to treat the site.

The second important human-made fuel complex is a special class of slash fuels created by unique fuel management activities that are becoming common on fire-prone landscapes of the world. The *masticated fuelbed* is created as a result of mechanical treatments that break canopy fuels and large surface fuels into smaller particles in an effort to decrease fire hazard by reducing canopy bulk densities and heights and decreasing surface fuel depths (Fig. 3.3f). Properties of the masticated fuel particles and fuel beds are unlike any observed in nature or documented in the past in that particles remaining after mastication treatments are shards of wood of odd shapes and sizes depending on the technique used to fragment the canopy and surface fuels (Kane et al. 2009). Diverse mechanical equipment and approaches are used to flail, chip, chop, and crush live and dead biomass to create a compact masticated fuel bed that many believe will burn slower and with lower intensities than pre-masticated fuel beds. However, current research has shown that mastication treatments do not always achieve fire management objectives (Ulanova 2000) and fires that burn in masticated fuels may be more intense, kill more trees, and cause more damage to soil biota than fires that burn in pre-masticated fuels.

Another similar special fuel type is stumps (Fig. 3.3e). Stumps are now a common occurrence on mechanically treated lands (Fule et al. 2001) and the biomass in stumps may contribute to many ecological and fire management concerns, such as smoke, wildlife habitat, and soil heating. Tree-cutting treatments, for example, may leave an abundance of stumps of various sizes, shapes, and species. Stumps, like snags, rarely burn in prescribed fires or wildfires unless they are thoroughly decomposed, yet they play important ecological roles such as stabilizing soil, providing long-term nutrient retention, and fostering biodiversity.

3.3 Ground Fuel Layer

The most common ground fuel type is duff and it is usually present as a result of the decay of the surface fuels. Most non-duff ground fuel layers are found in special locations where organic material may accumulate because of retarded decomposition or heavy biomass deposition.

3.3.1 Duff

Duff is the layer of decomposing organic materials lying just below the litter layer and immediately above the mineral soil (Fig. 3.1a). However, this simple definition is difficult to employ in the field because of the pronounced gradient of decomposition from the top of the litter layer (freshly fallen material) to the mineral soil (most decomposed material). Therefore, a better, operational definition of duff is the decomposing layer of biomass where the original source of fuel particles is no longer identifiable above mineral soil and below the litter layer. In reality, most dead biomass in the duff is in some stage of decomposition and this contributes to a corresponding wide range of duff fuel properties, such as moisture, mineral content, heat content, combustion efficiency, and smoldering duration across the duff profile. To remedy this, some have divided the duff layer into the upper and lower duff (Brown et al. 1991). The *upper duff* is usually referred to as the *fermentation* layer (F layer) where some fuel particle origins are still somewhat identifiable. The *lower duff* is often called the *humus* layer (H layer) where the particles are too decomposed to be identified (Soil Classification Working Group 1998). The Soil Survey Staff (2006) have defined the upper duff as the O_e layer and lower duff as the O_a layer. However, this upper and lower duff differentiation is also difficult to employ in the field in that it is nearly impossible to consistently identify where the upper duff ends and lower duff starts. Here, duff is considered one layer fully recognizing the great amount of variability in fuel properties across the thickness of this component.

Duff properties are especially difficult to define and measure because of other factors besides the decomposition gradient. The novel fuel beds created by new fuel treatments provide an example. Mastication, the modification of the fuel bed through mechanical means, often creates fuel particles whose origins are also difficult to identify but they are in the early stages of decomposition. So should these masticated materials be considered part of the duff if they are distributed throughout the duff profile? Another aspect of duff that is difficult to evaluate in the field is whether a decomposed piece of wood is considered duff or woody fuels. Some logs are so decomposed that the remaining biomass often burns more like duff than woody material, yet the features of the log are still fully recognizable. When you kick these logs, they often break apart with little effort. In this book, rotten logs are duff only if the central axis of the log (parallel to log length) is below the top of the litter (Lutes et al. 2006). This physical definition is a typical example of the differences between biomass and fuel, and the difficulty in operationally defining biomass as fuel. Sound woody fuels often protrude into the duff and are easily recognizable in the duff profile, yet operationally, these woody fuels are often considered part of the duff; only those woody particles above the duff and litter layers are considered woody fuels. These examples illustrate how difficult it is to measure and understand duff dynamics. There is considerable overlap between fuel types, and then, once fuel types are defined as components, there is a smaller, but still considerable, overlap between fuel components. It is nearly impossible to stratify fuel components so that they are always easily and accurately identified in the field.

Duff is important to fire behavior and effects in a way that is quite different from surface fuels. Duff necromass rarely contributes to fire spread because, in general, it is densely packed so it dries slowly and is often moister than other fuel components making it difficult to ignite and burn. It also consists of highly decomposed material just above the mineral soil so it usually has a high mineral content that dampens fire spread. As a result, duff layers burn mostly under smoldering rather than flaming combustion, although the upper duff may contribute to a surface fire if it is dry enough and there are sufficient litter fuels.

Duff has major implications for fire management because most of the dead necromass in a fuel bed is usually contained in the duff and logs (Table 3.2). It is common for logs and duff to contain over 60 % of the surface fuel loading in most forested ecosystems (Brown and Bevins 1986), and duff often comprises the majority of that 60 % because of its high bulk density values that range from 12 kg m^{-3} in deciduous ecosystems to over 150 kg m^{-3} in decomposed logs in duff of conifer ecosystems (Brown 1981; Stephens et al. 2004). And, because of high moisture and mineral content, duff biomass burns produces great amounts of smoke during smoldering combustion that can reduce visibility, lower air quality, and adversely impact the health of people.

Table 3.2 Loadings (kg m^{-2}) of each of the eight primary fuel components for the surface and ground fuel layers (Table 3.1) along with a measure of variation (coefficient of variation is in parentheses %) for selected undisturbed shrub and pine fuel beds in the USA. These loadings were taken from default inputs in the First Order Fire Effects Model (FOFEM), (Reinhardt et al. 1997), fuel characteristics classification system (Ottmar et al. 2007), and Keane et al. (2012), while the uncertainty estimates (in parentheses) were estimated from Keane et al. (2012). Coefficient of variation is the variation expressed as percent of the mean

Fuel comp	Low sagebrush	Chamise chaparral	Pinyon-juniper	Ponderosa pine	Jeffrey pine	Loblolly pine	Lodgepole pine
1 h	0.00 (0)	0.01 (60)	0.01 (110)	0.02 (80)	0.02 (80)	0.02 (75)	0.07 (86)
10 h	0.01 (200)	0.01 (200)	0.02 (300)	0.08 (193)	0.18 (193)	0.0 (0)	0.14 (131)
100 h	0.00 (0)	0.01 (140)	0.02 (440)	0.08 (185)	0.22 (185)	0.0 (0)	0.05 (253)
1000 h	0.00 (0)	0.0 (600)	0.11 (600)	0.66 (105)	2.21 (105)	0.0 (0)	0.59 (125)
Shrub	0.22 (50)	2.97 (202)	0.09 (202)	0.05 (177)	0.04 (177)	0.30 (150)	0.05 (78)
Herb	0.04 (251)	0.01 (850)	0.01 (850)	0.06 (44)	0.04 (44)	0.02 (102)	0.03 (127)
Litter	0.02 (110)	0.01 (100)	0.20 (100)	0.63 (58)	0.33 (58)	0.21 (103)	0.51 (64)
Duff	0.19 (105)	0.23 (150)	1.23 (150)	1.90 (55)	5.50 (55)	0.97 (120)	4.02 (64)

3.3.2 *Special Ground Fuel Types*

Plant roots are a ground fuel type that is often ignored in many fuel sampling efforts. Roots are relatively unimportant to fire spread and intensity because of their heterogeneous spatial distributions and position below the ground. However, fires will long persist and travel in dead roots causing major problems in fire suppression efforts. Smoldering root fires may be difficult to detect and may eventually transition into a surface fire given the right environmental conditions. There are few operational or research sampling methods to quantify root biomass, and fewer studies have described root combustion properties. Many recognize the great amount of carbon from roots in the soil profile.

Another important ground fuel type includes the diverse organic soils of the world's wetlands (Fig. 3.3g). Wetlands are fresh, saline, or brackish water bodies where biomass accumulates over time to create large deposits of organic soils. The organic matter at the bottom of the wetland better fit the definition of soil rather than fuel because of their high mineral contents, but many of these wetlands experience seasonal drying and often burn. These wetlands include the pocosin swamps of the southeastern USA; the boreal peat soils and bogs of Canada, USA, and Russia; the freshwater marshes of China; and the peatlands of Indonesia. In soils terminology, these soils are often called histosols because of high mineral contents and thick histic (organic) epipedons. Histosols are wide spread across various wetland settings, but there are many other soil types that also have high organic fuels that do not fully meet the classification requirements of a histosol. Gelisols, for example, are associated with areas of permafrost and have significant organic accumulation, but they do not meet the definition of a histosol. However, many gelisols burn when dry. Soils not deep enough to meet the histosol requirements may still have a thick organic soil horizon and may be considered organic soil fuels. Organic soils can be quite deep and have high organic matter loadings that, when burned, contribute tremendous amounts of smoke and carbon into the atmosphere, that have important implications for air quality, human health, and climate change.

References

Albini FA, Reinhardt ED (1995) Modeling ignition and burning rate of large woody natural fuels. Int J Wildland Fire 5(2):81–91

Anderson HE (1969) Heat transfer and fire spread. U.S. Department of Agriculture, Forest Service, Intermountain Forest and Range Experiment Station, Research Paper INT-69. Ogden, UT, USA, 20 pp

Andreu AG, Shea D, Parresol BR, Ottmar RD (2012) Evaluating fuel complexes for fire hazard mitigation planning in the southeastern United States. For Ecol Manage 273(0):4–16. doi:http://dx.doi.org/10.1016/j.foreco.2011.06.040

Brown JK (1981) Bulk densities of nonuniform surface fuels and their application to fire modeling. For Sci 27:667–683

Brown JK, Bevins CD (1986) Surface fuel loadings and predicted fire behavior for vegetation types in the northern Rocky Mountains. U.S. Department of Agriculture, Forest Service, Intermountain Forest and Range Experiment Station, Research Paper INT-358. Ogden, UT USA, 9 pp

Brown JK, Oberheu RD, Johnston CM (1982) Handbook for inventoring surface fuels and biomass in the Interior West. USDA Forest Service, Intermountain Forest and Range Experiment Station, General Technical Report INT-129. Ogden, UT, USA, 48 pp

Brown JK, Reinhardt ED, Fischer WC (1991) Predicting duff and woody fuel consumption in northern Idaho prescribed fires. For Sci 37(6):1550–1566

Busing RT, White RD, Harmon ME, White PS (2009) Hurricane disturbance in a temperate deciduous forest: patch dynamics, tree mortality, and coarse woody detritus. Plant Ecol 201(1):351–363. doi:10.1007/s11258-008-9520-0

Flavelle M (2008) Kimberely grasslands field curing guide. Kimberely Regional Fire Management Project, Broome. CSIRO Publishing, Western Australia

Fosberg MA, Lancaster JW, Schroeder MJ (1970) Fuel moisture response–drying relationships under standard and field conditions. For Sci 16(1):121–128

Frandsen WH (1983) Modeling big sagebrush as a fuel. J Range Manage 36(5):596–600

Fule PZ, McHugh C, Heinlein TA, Covington WW (2001) Potential fire behavior is reduced following forest restoration treatments. In: Regina K, Edminster CB, Covington WW, Blake JA (eds) Ponderosa pine ecosystems restoration and conservation: steps toward stewardship, Flagstaff, pp 28–35. USDA Forest Service Rocky Mountain Research Station, Proceedings RMRS-P-22

Harmon ME, Franklin JF, Swanson FJ, Sollins P, Gregory SV, Lattin JD, Anderson NH, Cline SP, Aumen NG, Sedell JR, Lienkaemper GW, Cromack K, Cummins KW (1986) Ecology of coarse woody debris in temperate ecosystems. Adv Ecol Res 15:133–302

Higgins SI, Bond WJ, Trollope WW (2000) Fire, resprouting and variability: a recipe for grass-tree coexistence in savanna. J Ecol 88:213–229

Hutto RL (2006) Toward meaningful snag-management guidelines for postfire salvage logging in North American conifer forests. Conserv Biol 20(4):984–993. doi:10.1111/j.1523-1739.2006.00494.x

Jenkins MA, Hebertson E, Page W, Jorgensen CA (2008) Bark beetles, fuels, fire and implications for forest management in the Intermountain West. For Ecol Manage 254:16–34

Kane JM, Varner JM, Knapp EE (2009) Novel fuel characteristics associated with mechanical mastication treatments in northern California and south-western Oregon, USA. Int J Wildland Fire 18:686–697

Keane RE, Dickinson LJ (2007) The Photoload sampling technique: estimating surface fuel loadings using downward looking photographs. USDA Forest Service Rocky Mountain Research Station General Technical Report RMRS-GTR-190, 44 pp

Keane RE (2008) Surface fuel litterfall and decomposition in the northern Rocky Mountains, USA. USDA Forest Service Rocky Mountain Research Station, Research Paper RMRS-RP-70. Fort Collins, Colorado USA, 22 pp

Keane R, Gray K, Bacciu V, Leirfallom S (2012) Spatial scaling of wildland fuels for six forest and rangeland ecosystems of the northern Rocky Mountains, USA. Landscape Ecol 27(8):1213–1234. doi:10.1007/s10980-012-9773-9

Keane RE, Gray K (2013) Comparing three sampling techniques for estimating fine woody down dead biomass. Int J Wildland Fire 22(8):1093–1107. doi: http://dx.doi.org/10.1071/WF13038

Lutes DC, Keane RE, Caratti JF, Key CH, Benson NC, Sutherland S, Gangi LJ (2006) FIREMON: Fire effects monitoring and inventory system. USDA Forest Service Rocky Mountain Research Station, General Technical Report RMRS-GTR-164-CD. Fort Collins, CO USA

Lutes DC, Keane RE, Caratti JF (2009) A surface fuels classification for estimating fire effects. Int J Wildland Fire 18:802–814

Nappi A, Drapeau P, Savard JPL (2004) Salvage logging after wildfire in the boreal forest: is it becoming a hot issue for wildlife? For Chron 80(1):67–74. doi:10.5558/tfc80067-1

Ottmar RD, Sandberg DV, Riccardi CL, Prichard SJ (2007) An overview of the fuel characteristic classification system—quantifying, classifying, and creating fuelbeds for resource planning. Can J Forest Res 37:2383–2393

Pyne SJ, Andrews PL, Laven RD (1996) Introduction to wildland fire, 2nd edn. Wiley, New York

Reinhardt E, Keane RE, Brown JK (1997) First order fire effects model: FOFEM 4.0 user's guide. USDA Forest Service, Intermountain Research Station General Technical Report INTGTR-344, 65 pp

Riccardi CL, Ottmar RD, Sandberg DV, Andreu A, Elman E, Kopper K, Long J (2007a) The fuelbed: a key element of the fuel characteristic classification system. Can J For Res 37:2394–2412

Riccardi CL, Prichard SJ, Sandberg DV, Ottmar RD (2007b) Quantifying physical characteristics of wildland fuels using the Fuel Characteristic Classification System. Can J For Res 37:2413–2420

Rothermel RC (1993) Mann Gulch fire: a race that couldn't be won. U.S. Department of Agriculture, Forest Service, Intermountain Research Station, General Technical Report INT-299. Ogden, UT USA, 10 pp

Shetler G, Turetsky MR, Kane E, Kasischke E (2008) Sphagnum mosses limit total carbon consumption during fire in Alaskan black spruce forests. Can J For Res 38(8):2328–2336. doi:10.1139/x08–057

Soil Classification Working Group (1998) The Canadian System of Soil Classification. Agriculture and Agri-Food Canada, Soil Classification Working Group, 3rd ed. The Canadian System of Soil Classification, 187 p

Soil Survey Staff (2006) Keys to Soil Taxonomy, 10th ed. U.S. Department of Agriculture Natural Resource Conservation Service, Washington, DC, USA

Stephens SL, Finney M, Schentz H (2004) Bulk density and fuel loads of ponderosa pine and white fir forest floors: Impacts of leaf morphology. Norhtwest Sci 78(2):93–104

Sylvester TW, Wein RW (1981) Fuel characteristics of arctic plant species and simulated plant community flammability by Rothermel's model. Can J Bot 59(5):898–907. doi:10.1139/b81-125

Ulanova NG (2000) The effects of windthrow on forests at different spatial scales: a review. For Ecol Manage 135(1–3):155–167. doi:http://dx.doi.org/10.1016/S0378-1127(00)00307-8

Woodall CW, Nagel LM (2007) Downed woody fuel loading dynamics of a large-scale blowdown in northern Minnesota, U.S.A. For Ecol Manage 247(1–3):194–199 doi:http://dx.doi.org/10.1016/j.foreco.2007.04.040

Chapter 4
Canopy Fuels

A forest of these trees is a spectacle too much for one man to see
David Douglas Scottish Botanist

4.1 Introduction

The forest canopy is an amazing ecological entity consisting of a wide variety of suspended and interacting plants, insects, pathogens, and many other life-forms, essentially creating an aerial ecosystem within a terrestrial ecosystem (Nadkarni 1994). The canopy habitat provides important food, shelter, and shade for critical wildlife species. It consists of a wide variety of life-forms, species, and plants (e.g., mosses, lichens, epiphytes, tree crowns) that have a wide variety of sizes, shapes, and other biophysical characteristics (e.g., foliage, twigs, branches, and boles). Interestingly, foliage is only a small fraction of the total canopy biomass; the woody material in tree boles accounts for over 80 % of the total canopy biomass (Table 4.1). Moreover, the forest canopy is constantly changing as climate, disturbance, and vegetation development processes interact to create unique tree crown and forest canopy mosaics across the landscape. Yet, even with this remarkable diversity of canopy species, structures, and conditions, wildland fuel science has described canopy biomass using only a handful of variables that relate poorly to many ecological processes. This chapter describes how canopy fuels are described, measured, and summarized for fire management and presents possible limitations of these approaches.

The concept of scale is again needed to fully understand current descriptions and quantifications of canopy fuels. Early studies often called forest aerial biomass "*crown*" fuels (van Wagner 1977). A crown is defined as the portion of a tree's height that is composed of branches that support live foliage. However, tree crowns may also have arboreal lichens and mosses, dead branches, and abnormal growths, such as split boles and dwarf mistletoe infections. The height to the bottom of the crown (HBC) and the height to the top of the crown (tree height; HT) (Fig. 4.1) are often used to compute canopy fuel vertical distribution. The problem is that tree crown characteristics describe fuels at the fine-scale tree-level, not at the broader canopy or stand-level, which is the scale that best matches crown fire spread and fuels management. Therefore, the term "*canopy*" fuels was introduced to capture

R. E. Keane, *Wildland Fuel Fundamentals and Applications*,
DOI 10.1007/978-3-319-09015-3_4

Table 4.1 Important canopy fuel characteristics for the four US northern Rocky Mountain stand types and one Sierran mixed conifer type sampled by Scott and Reinhardt (2005). *CBH* is canopy base height, *CH* is canopy height, *CBD* is canopy bulk density, and *CFL* is canopy fuel load

Canopy fuel variable	Douglas-fir lodgepole pine	Lodgepole pine	Ponderosa pine douglas-fir	Ponderosa pine	Sierran mixed conifer
CBH (m)	1.0	1.0	0.0	5.0	3.0
CH (m)	17.0	20.0	23.0	16.0	34.0
CBD (kg m^{-3})	0.26	0.11	0.09	0.17	0.10
CFL (kg m^{-2})	2.09	1.00	1.40	0.93	1.72
Foliage biomass (kg m^{-2})	1.59	0.80	1.11	0.88	1.48
Branch biomass (kg m^{-2})	3.26	2.77	3.18	3.66	3.80
Bole biomass (kg m^{-2})	11.22	13.88	10.08	16.53	18.61
Total aboveground biomass (kg m^{-2})	16.07	17.45	14.36	21.07	23.89

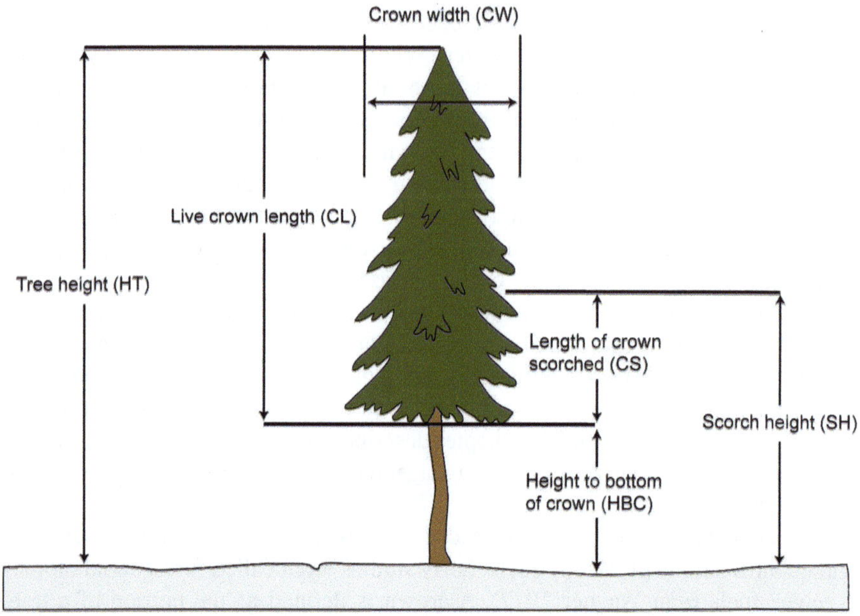

Fig. 4.1 The dimensions of a tree that are important in describing and calculating canopy fuel characteristics

the description of crown fuels at scales more appropriate for fire modeling. Canopy fuels are often described using a summary of crown fuels across all trees in an area from plot- or stand-level measurement techniques (Keane et al. 2005) (Chap. 8). One confusing conflict in terminology is fires that spread in the canopy are often called "crown" fires, not canopy fires. And to add to the confusion, when crown fires occur, they can burn at various scales; passive crown fires often burn at the tree- and branch-scale, whereas active crown fires may burn at the canopy-scale or maybe even coarser scales of hillsides or topographic settings. Canopy fuels have their own intrinsic scale for description and measurement, and this scale is coarser than surface fuels and finer than active crown fires (Keane et al. 2012a). This becomes important when surface fuels are sampled at the same time as canopy fuels (Chap. 8).

Canopy fuels are probably the most misunderstood and misrepresented fuel type in wildland fuel science. The reason for this is that most fire behavior models simulate crown fire behavior for operational fire management using a simplistic approach (Rothermel 1991; Finney 1998). As a result, there really hasn't been the need to stratify the complex forest canopy fuel properties by various fuel particles and components. Therefore, to fully understand the contemporary characterizations of canopy fuels, it is important to understand how crown fires are currently being simulated.

4.2 Crown Fire Simulation

In a first attempt to describe canopy fuels for operational use for rating fire hazard, Fahnestock (1970) created a key to rate crowning potential based on canopy closure, crown density, and ladder fuels. While effective, a more mechanistic, detailed approach was needed for robust applications at diverse local conditions. The first operational attempt at quantitative crown fire modeling was by van Wagner (1977) who developed a mathematical classification system to determine if a surface fire transitions to a crown fire, and then, once the fire was in the canopy, whether the fire actively spread through the crown or the crown fire was passive and merely torched individual trees. This basic model has since been modified (van Wagner 1993) and it has also been merged with other models (Scott and Reinhardt 2001) for implementation in various fire behavior prediction systems, such as FARSITE (Finney 1998). Material presented here were synthesized from several sources (Cruz and Alexander 2010; Finney 1998; Alexander 1998; Scott and Reinhardt 2001).

In general, crown fire behavior is simulated by comparing fire intensity (*I* variables below) and spread rates (*R* variables) to critical crown fire thresholds. The van Wagner (1993) crown fire model simulates the intensity threshold for transition to crown fire I_o (kW m^{-1}) from crown foliar moisture content (*FMC*, % dry weight) and canopy base height (*CBH*, m) (van Wagner 1977; van Wagner 1993; Nabel et al. 2014) using the following empirical relationship:

$$I_o = 0.01CBH(460 + 25.9FMC)^{1.5} \tag{4.1}$$

CBH was defined as the vertical distance between the ground surface and the base of the live crown fuels, but, many have advocated that it should be expanded to the canopy-scale instead of an individual tree so that it may include all types of "ladder" fuels including small trees, large shrubs, and attached dead branches (Alexander 1988; FCFDG 1992). The model keys the type of crown fire depending on the threshold for active crown fire spread rate *RAC* (Alexander 1988):

$$RAC = \frac{3.0}{CBD} \tag{4.2}$$

where *CBD* is the canopy bulk density (kg m^{-3}) and the 3.0 coefficient is an empirical constant defining the critical mass flow rate through the crown layer for continuous flame (0.05 kg m^{-2} s^{-1}) and a conversion factor (60 s min^{-1}). Van Wagner (1977) estimated *CBD* by computing canopy biomass (kg) of all trees in a unit area and dividing it by the canopy volume (unit area times the canopy depth defined as the canopy height - CH, m - minus *CBH*). One of three types of crown fire (passive, active, and independent) is then keyed using the I_o and *RAC* values. A passive crown fire must meet the following conditions:

$$I_b = I_o, \text{ but } R_{C\,\text{actual}} < RAC \tag{4.3}$$

where I_b is the surface fireline intensity (see Chap. 2.2) and $R_{C\,\text{actual}}$ is the active crown fire spread rate (m min^{-1}) calculated from the following using the FCFDG (1992) approach:

$$R_{C\,\text{actual}} = R + CFB(R_{C\,\max} - R) \tag{4.4}$$

where *R* is the rate of spread (m min^{-1}) of the surface fire (see Chap. 2.2) and *CFB* is crown fraction burned, a scalar from 0 to 0.9 computed from RAC and a critical surface fire spread rate (R_o) (Finney 1998; Scott and Reinhardt 2001). The variable $R_{C\,max}$ is the maximum crown fire spread rate (m min^{-1}) that many estimate using the Rothermel (1991) algorithm where $R_{C\,max}$ is computed from a correlation with the forward surface fire spread rate for the Anderson (1982) US fuel model 10 using a 0.4 wind reduction factor (R_{10} m min^{-1}) from the following relationship:

$$R_{C\max} = 3.34 R_{10} \tag{4.5}$$

The rate of spread of a passive crown fire is assumed to be equal to that of the surface fire.

If passive conditions aren't met, then conditions for active crown fire must meet the following criteria:

$$I_b \geq I_o \text{ and } R_{C\,\text{actual}} \geq RAC \tag{4.6}$$

If conditions for active or passive crown fire are not met, then an independent crown fire is simulated. Linking the van Wagner (1977); Nabel et al. (2014); and van Wagner (1993) crown fire algorithms with Rothermel (1972) and Rothermel (1993) surface fire algorithms required many assumptions, modifications, and scaling factors for operational use (Finney 1998).

The final intensity of a crown fire I_c (kW m^{-1}) is calculated using the combined loading of both canopy and surface fuel consumed in the flaming front along with the rate of spread for active crown fires ($R_{C\,\mathrm{actual}}$) or for passive crown fires (surface fire spread rate R):

$$I_c = 300\left[\frac{I_b}{300R} + (CFB)(CBD)(CH - CBH)\right]R_{\mathrm{C\ actual}} \tag{4.7}$$

where CH is canopy height (m), CFB is canopy fuel burned (kg m^{-2}) estimated from the canopy fuel load (CFL, kg^{-2}) assuming all is consumed, and R is substituted for $R_{C\,\mathrm{actual}}$ if a passive crown fire.

Several limitations of these algorithms need mention. First, the basic relationships in this model are empirical functions mostly derived from a limited set of vegetation types (van Wagner 1977). There has been little validation of this model for many fire-prone forested vegetation types of the world because of the difficulty of experimentation with crown fires (e.g., Stocks et al. 2004). Next, the quantitative linkages between surface and crown fire simulation are difficult to implement because so many assumptions on the scale, behavior, and inertia of the fire had to be made and these assumptions often introduced additional uncertainty in crown fire behavior.

It is also important to note that the above crown fire model formulation only concerns operational fire models. More sophisticated physics-based CFD (computational fluid dynamics) modeling approaches are making improvements in the simulation of crown fires (Linn 1997; Parsons et al. 2010), but there are still limitations of this approach (Alexander and Cruz 2013; Zhang et al. 2014). CFD models simulate the behavior of a fire in three dimensions so fuels must be described in three-dimensional volumetric cells called "voxels." Parsons (2006) developed a tool called FUEL3D that maps canopy fuel into voxels using allometric and fractal relationships of tree biomass to simulate crown fires in CFD models. While CFD modeling is mostly in the research stage with limited evaluation and most models haven't yet been implemented for operational use, they illustrate the importance of describing canopy fuels in a three-dimensional context.

From this thumbnail summary of operational crown fire simulation, it appears that there are four variables in crown fire simulations that represent canopy fuels: *CBH, CBD, CH,* and *CFL*. Crown fire initiation is dependent on *CBH* and *CBD* (Eq. 5.1), while crown fire spread and intensity are dependent on *CBD, CH,* and *CFL* (Eqs. 5.2, 5.7). One other canopy characteristic, canopy cover (*CC*, %), is also discussed here along with *CBD, CH, CBH,* and *CFL* because it is important in fuel moisture and fire effects simulation.

4.3 Canopy Fuel Characteristics

Given the simplistic empirical approach in which canopy fuels are represented in the equations above, canopy fuels did not need the complex array of fuel components used for surface fire simulation. A problem arose, however, when many observed that only the smaller canopy fuels, such as needles and small branches, actually burned in crown fires, and that inclusion of the larger canopy fuel (branches, tree boles) was perhaps inappropriate for crown fire modeling (Call and Albini 1997). Indeed, loadings of these smaller canopy materials are significantly less than unconsumed loadings of the larger branches and boles (Table 4.1). So, most canopy fuel descriptions today are quantified using only "burnable" canopy biomass or canopy fuel available for combustion.

The canopy fuel particle size range that defines burnable canopy biomass has been described in many studies. Brown and Reinhardt (1991) suggested estimating burnable canopy fuels as 50 % of the crown branchwood that is less than 6 mm in diameter and all of foliar material. Reinhardt et al. (2006b) specified burnable canopy biomass as all foliage and live branchwood less than 3 mm diameter, and dead branchwood less than 6 mm, while Keane et al. (2005) defined canopy fuels as all crown fuels less than 3 mm diameter. Call and Albini (1997) estimated that 65 % of the canopy biomass less than 6 mm burned in a crown fire. Stocks et al. (2004) conducted one of the few studies that actually measured canopy fuel consumption and found nearly all canopy material less than 1.0 cm were consumed in a boreal crown fires. Obviously, the amount of burnable canopy biomass differs by ecosystem, time of year, weather, and fire, so defining the size for burnable canopy material would depend on the analysis objectives. For most applications, the three canopy fuel characteristics needed for crown fire behavior prediction (*CBD, CH,* and *CBH*) are calculated using only burnable canopy biomass, mainly so that *CFB* equals *CFL* in Eq. 4.7. However, to fully understand how the three canopy fuel characteristics are quantified, it is important to know how canopy fuel is arrayed above the ground using vertical distributions of canopy fuels called canopy profiles.

4.3.1 The Canopy Fuel Profile

The canopy fuel profile is the vertical distribution of burnable canopy biomass above the ground (Fig. 4.2). For operational uses, this can be envisioned by slicing the canopy into a numerous horizontal layers, and for each layer, summing all the canopy fuel weights within each layer and dividing by the volume of that layer to estimate a *CBD* for each layer. Plotting the *CBD* for each layer by the height of each layer gives the canopy fuel profile in Fig. 4.2. Canopy fuel profiles can be measured directly using destructive sampling or they can be modeled using allometric tree biomass estimates and the tree dimensions in Fig. 4.1 (Reinhardt et al. 2006b). Canopy fuel profiles can be generated for both loading (*CFL*) and bulk density (*CBD*), but here *CBD* profiles are used because they better represent inputs into the fire models.

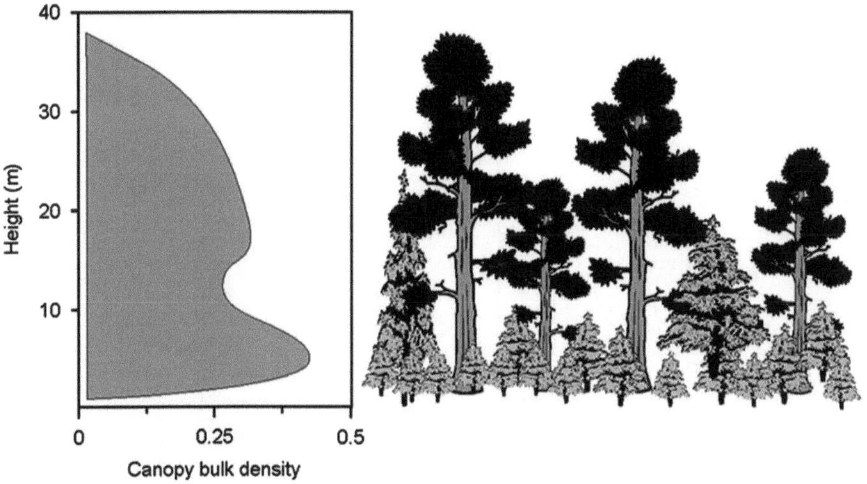

Fig. 4.2 Example of a typical canopy profile where canopy bulk density (*CBD*, kg m^{-3}) is plotted across the vertical profile of a stand. (From Reinhardt et al. 2006b)

There are four basic canopy *CBD* profiles that generally describe the potential for crown fires. Multiage, multistrata stands (Fig. 4.3a) have trees at a wide variety of heights with mostly shade-tolerant trees comprising the understory so canopy fuels are dense even at the surface-canopy fuel interface creating a high potential for crown fires because of the low *CBH* and high *CBD*. Two storied stands (Fig. 4.3b), commonly having a shade-intolerant, fire-resistant overstory with a shade-tolerant, fire-susceptible understory, have hourglass-shaped canopy profiles that also have high *CBD* and low *CBH*. These are often the profiles of stands that had frequent fires but recent fire exclusion has allowed increases in understory biomass. Single-strata stands have profiles where the majority of biomass is at the top of the canopy creating a low chance of crown fire initiation (high *CBH*) but a high conditional probability of crown fire spread (high *CBD*) (Fig. 4.3c). The last profile is a forest stand with trees that are widely spaced and somewhat open so that the *CBD* rarely exceeds the threshold for crown fire initiation and spread creating a low chance of both initiation and spread of a crown fire (Fig. 4.3d). This is a typical canopy profile for a stand with a history of frequent fires, woodland, or perhaps a stand with recent insect- or disease-related tree mortality.

4.3.2 Canopy Bulk Density

As canopy fuel characteristics go, *CBD* is probably the most important because it dictates crown fire spread, potential for an active crown fire, and crown fire intensity (Eqs. 4.2, 4.7). *CBD* is the amount of burnable canopy fuel by canopy volume (kg m^{-3}) and represents the degree of packing of canopy fuels. Below a minimum *CBD* threshold value, crown fires have insufficient fuels for crown fire spread.

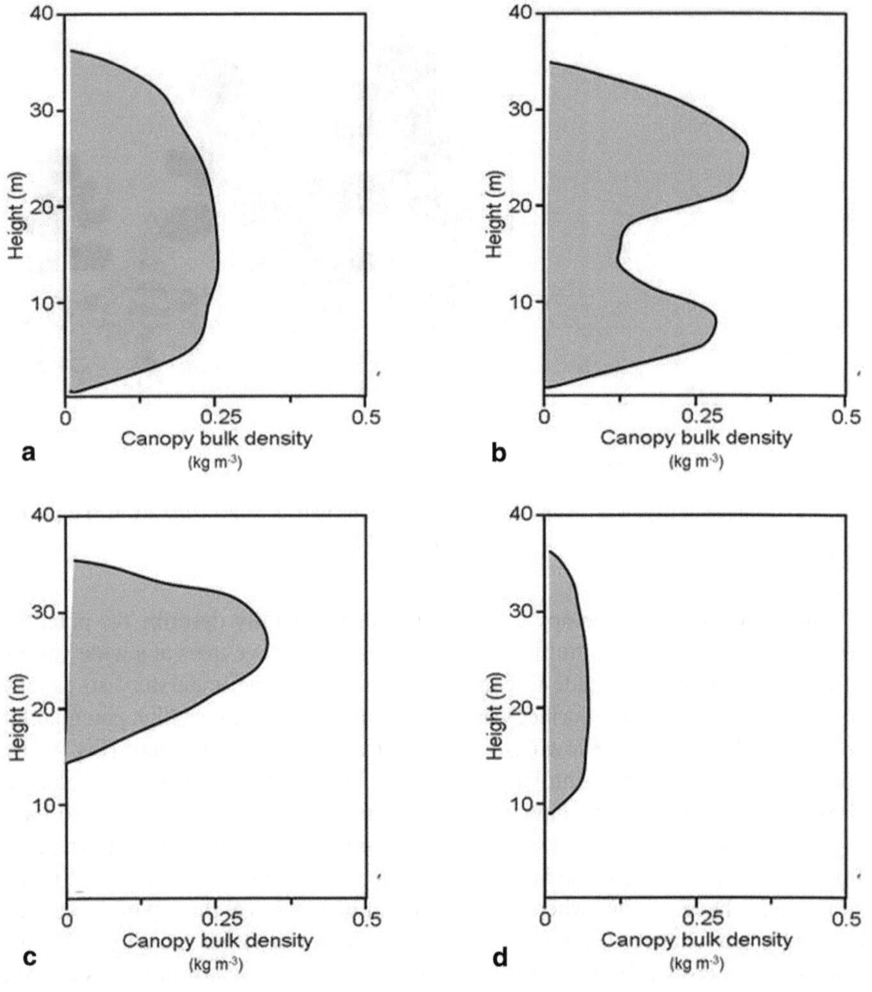

Fig. 4.3 Four typical canopy fuel profiles. **a** multiaged, multistrata stands, **b** two storied, **c** single strata, and **d** fire-dominated stands. Each stand presents a special problem in calculating *CBD* and *CBH*

There is also a maximum threshold, but this is rarely approached in nature. *CBD* values can range from zero for grasslands to over 0.7 kg m^{-3} for the densest canopies (Reeves et al. 2009) (Table 4.1).

The two main problems with *CBD* are how to define it and how to estimate it. Cruz et al. (2003) defined *CBD* following the van Wagner (1977) approach where canopy fuel load (*CFL*, kg m^{-2}) is divided by canopy depth (*CH–CBH*, m), but this average *CBD* across all canopy layers doesn't really capture the true nature of the canopy profile; some canopy layers are denser thereby better facilitating the spread of crown fires (Fig. 4.2). However, selecting the maximum *CBD* value across all layers ignores the role that other canopy layers play in crown fire spread (Keane et al. 2005) and may overestimate crown fire behavior in some canopies that have unique profiles because it may be incompatible with the original van Wagner (1977) model assumptions (Fig. 4.2a). Early attempts at rectifying this problem calculated *CBD* using a running mean of *CBD* across all canopy layers with the number of layers included in the running mean ranging from 3 (one below and one above) to 11 (five layers below and five layers above the evaluated layer) (Reinhardt et al. 2006a, b).

A limitation of these averaging techniques is that the mean *CBD* value often masked the great variability of canopy fuel over horizontal space (Keane et al. 2012b). Most *CBD* estimates are derived from tree inventories summarized from plots sampled across large areas disregarding the actual locations of trees that contribute to the heterogeneous distribution of canopy fuels (a summary of all canopy fuel sampling techniques are detailed in Chap. 7). A stand-level summary of *CBD* may be too coarse for accurately representing the spatial distribution of *CBD* at the scales that fires spread. On the other hand, plot-level summaries of *CBD* may match the scale of crown fire behavior, but assigning one plot to represent canopy fuel conditions across an entire stand also ignores the high spatial variability of canopy fuels. There may be canopy patches that are unable to sustain crown fire spread, and the height of the densest *CBD* layers often varies across heterogeneous forests.

An accurate, scale-appropriate, and physically credible estimate of *CBD* is probably not a great concern right now given the coarse resolution of the current crown fire modeling approaches (Eq. 4.1). Users of the FARSITE model and its derivatives (Finney 1998) often adjust the carefully field-calculated estimates of *CBD* inputs to more realistically simulate crown fire spread due to limitations in the design of the fire model (Keane et al. 2006; Reeves et al. 2006). And since *CBD* can be estimated using any number of techniques, its quantification can be modified to fit a particular application. The *CBD* variable was initially selected by van Wagner (1977, 1993) to represent "canopy opaqueness" and perhaps there are more ecologically appropriate and easily measured variables that better represent this vague physical concept at the appropriate scale and resolution, such as Leaf Area Index (LAI). Future crown fire simulation systems may also find that the canopy fuel characteristic *CBD* could be replaced, modified, or redefined to fit more comprehensive spatial simulations (Parsons et al. 2010).

4.3.3 Canopy Base Height and Canopy Height

CBH is most important in estimating the potential of surface fires to transition to crown fires (Eqs. 4.1, 4.7). *CH* is less important because it is only used to calculate crown fire intensity (Eq. 4.7). Similar to *CBD,* both *CBH* and *CH* are difficult canopy characteristics to describe and measure because of major conflicts in spatial scales. *CBH* is defined as the lowest height above the ground at which there is sufficient canopy fuel to propagate fire vertically through the canopy (Scott and Reinhardt 2002). The problem is that a single sapling may provide the vertical pathway or ladder fuel needed to move surface fire into the canopy under the right conditions, but since many forest stands have at least one sapling reaching the canopy, it probably isn't appropriate to evaluate *CBH* at the fine-scale tree-level. The spatial distribution of ladder fuel across an area is probably more important; clusters of saplings and pockets of dense, low-hanging canopies better facilitate crown fire transitions. However, few available management-oriented sampling techniques record tree location. The scale of the fire is also important. Heavy surface fuel loadings may foster intense fires that can easily reach the forest canopy via flame or embers without the need for low fuel ladders, and large running crown fires may initiate crown fire from adjacent stand making *CBH* irrelevant.

Many techniques have been used to estimate *CBH* and *CH*. Some studies estimated *CBH* and *CH* as an average of crown base height and tree height, respectively, across all trees in the sampling frame (Chap. 8). Most studies used the canopy profile to estimate these two canopy parameters where *CBH* is estimated as the lowest height that exceeds a threshold *CBH* value and *CH* is the highest height above a threshold *CBD* (Fig. 4.4). Threshold values are difficult to estimate since there has been little experimental research in this area, so most studies select a value that works best for a given application. Sando and Wick (1972) estimated *CBH* using a minimum *CBD* threshold of 0.037 kg m^{-3} whereas Reinhardt et al. (2006a) used 0.012 kg m^{-3}. Agee (1996) has suggested that a minimum canopy bulk density of 0.1 kg m^{-3} is needed to ensure active crown fire spread in a horizontal dimension. The newest version of FuelCalc (Reinhardt et al. 2006a), a software application

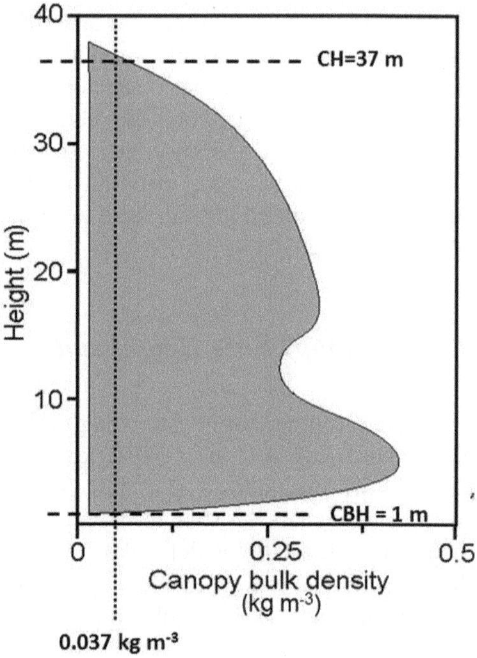

Fig. 4.4 The threshold method of estimating *CBH* and *CH* from the canopy profile. Here a threshold of 0.037 kg m^{-3} is used to detect the bottom (CBH) and top (CH)canopy heights at which CBD exceeds this threshold

specifically designed to estimate canopy fuel characteristics, computes both *CBH* and *CH* by multiplying the maximum *CBD* by 0.1 up to the *CBD* threshold of 0.012 kg m^{-3}. Since *CBH* and *CH* can be computed from so many diverse methods, it is important that those who use canopy fuel data to simulate fire behavior know how *CBH* and *CH* were calculated.

There are many aspects of the two canopy height estimates that may compromise their use in fire behavior analyses. First, regardless of what method is used to estimate *CBH* and *CH* from the modeled *CBD* canopy profile, the resolution of that estimate is governed by the width of the horizontal canopy layer used to summarize *CBH* vertical distribution. This is somewhat of a double-edged sword in that thin layer widths (high resolution) rarely match the resolution of the tree input data used to calculate the *CBD* estimates, but, the thicker layer widths that better match the scale of the data (> 1 m) may be too coarse to recognize subtle differences created by fuel treatments. Input tree data often contain only heights of the tree and crown base, and rarely have finer crown measurement resolutions. Second, surface fuels may have more importance in crown fire initiation than low *CBH* values. A small fire can ignite a canopy if there are sufficient ladder fuels at a tree-level, and a large fire can ignite a canopy no matter the *CBH.* And lastly and most importantly, the scale of *CBH* and *CH* measurements do not match the scale of the point-scale simulation models that simulate these processes, similar to *CBD.* The coarse resolutions of point-level, crown fire simulation in most fire behavior prediction systems is often incompatible with the fine-scale spatial distribution of *CBH* and *CH,* and the fine scales at which fire spreads. As a result, one can view *CBH* and *CH* as indices rather than actual measurements, and it is often a common practice to adjust the calculated or measured values to values that represent more realistic fire behavior.

4.3.4 Canopy Fuel Load

CFL (kg m^{-2}) is used to estimate the amount of canopy material consumed in a crown fire (*CFB* in Eq. 4.7), and it is defined as the amount of burnable canopy fuel per unit area. *CFL* is also quite useful in fire effects simulations, such as the estimation of smoke emissions (Ward 1995), because it provides a somewhat accurate representation of actual biomass consumed in a crown fire.

CFL isn't really related to any of the other three canopy characteristics (Table 4.1). While an estimate of *CFL* could be made by multiplying *CBD* by the difference between *CH* and *CBH,* this would assume an even distribution of *CBH* across the canopy fuel profile similar to van Wagner (1977) assumptions, but this is rarely observed as is evidenced by the profiles in Fig. 4.3 only a few methods calculate *CBD* as an average across the canopy profile (Cruz et al. 2003). FuelCalc, for example, estimates *CBD* as a maximum running average (Reinhardt et al. 2006a). Therefore, *CFL* needs to be estimated and mapped directly to ensure accurate and consistent fire applications.

CFL is also important because it may be linked to many other important ecosystem attributes and canopy processes, such as productivity and light attenuation (Keane et al. 2012b). LAI is an important canopy characteristic in forest ecology because it determines rates of photosynthesis, shading, and precipitation interception (rainfall that doesn't reach the ground) (Waring and Running 1998), and it may be highly related to *CFL* because *CFL* is mostly foliage (Keane et al. 2005; Scott and Reinhardt 2005) (Table 4.1). Like LAI, *CFL* is often correlated to common stand measurements, such as basal area, tree density, and stand height (Alexander and Cruz 2014). The relationship of *CFL* to stand attributes may allow the indirect mapping of *CFL* using maps of stand variables (LAI, basal area) or using a statistical model that relates *CFL* to the stand variables developed from field data (Chap. 9) (Cruz et al. 2003).

4.3.5 Canopy Cover

A last canopy fuel description variable that is needed outside of crown fire behavior prediction but is often used in fire behavior modeling is *CC*, or the vertically projected cover of the suspended canopy onto the ground. In many fire behavior modeling systems, canopy cover is used to simulate fuel drying in forested stands and, with *CH,* it is used to reduce midflame windspeeds (Finney 1998). Canopy cover is a canopy attribute that is used extensively in forestry and ecology, along with its application in fire science. It is difficult to estimate *CC* in field settings with any degree of accuracy or precision because of the wide diversity of canopy material and the heterogeneous vertical profile of the canopy. Therefore, many estimate *CC* to the nearest 10% using visual estimates or coarse measurement techniques (Chap. 8). It has all the scale and measurement problems of *CBD, CH,* and *CBH* in fire behavior and effects modeling.

References

Agee JK (1996) The influence of forest structure on fire behavior. In: Proceedings of the 17th annual forest vegetation management conference, Redding. CA, USA, January 16–18, pp 52–68

Alexander ME (1988) Help with making crown fire hazard assessments. In: Fischer W, Arno SF (eds) Protecting people and homes from wildfire in the interior West: proceedings of the symposium and workshop. General Technical Report INT-251. USDA Forest Service, Intermountain Research Station, Ogden, pp 147–153

Alexander ME (1998) Crown fire thresholds in exotic pine plantations of Australasia. Ph.D. dissertation, Australian National University. Canberra, Australia

Alexander ME, Cruz MG (2013) Are the applications of wildland fire behaviour models getting ahead of their evaluation again? Environ Model Softw 41:65–71. doi:http://dx.doi.org/10.1016/j.envsoft.2012.11.001

Alexander ME, Cruz MG (2014) Tables for estimating canopy fuel characteristics from stand variables in four Interior West conifer forest types. Forest Science 60(4)784–794

Anderson HE (1982) Aids to determining fuel models for estimating fire behavior. General technical report INT-122, USDA Forest Service Intermountain Research Station, Ogden, Utah, USA, 22 pp

Brown JK, Reinhardt E (1991) Estimating and regulating fuel consumption to manage smoke in the Interior West. In: Proceedings of the 11th conference on fire and forest meterology, Missoula, Montana. Society of American Foresters, Bethesda, MD, USA, pp 419–429

Call PT, Albini FA (1997) Aerial and surface fuel consumption in crown fires. Int J Wildland Fire 7(3):259–264

Cruz MG, Alexander ME (2010) Assessing crown fire potential in coniferous forests of western North America: a critique of current approaches and recent simulation studies. Int J Wildland Fire 19(4):377–398. doi:http://dx.doi.org/10.1071/WF08132

Cruz MG, Alexander ME, Wakimoto RH (2003) Assessing canopy fuel stratum characteristics in crown fire prone fuel types of western North America. Int J Wildland Fire 12:39–50

Fahnestock GB (1970) Two keys for appraising forest fire fuels. USDA Forest Service, Pacific Northwest Forest and Range Experiment Station, Res. Bull. PNW-RB-099, Portland, OR, 31 pp

Finney MA (1998) FARSITE: Fire Area Simulator—model development and evaluation. Research Paper RMRS-RP-4, United States Department of Agriculture, Forest Service Rocky Mountain Research Station, Fort Collins, CO, USA, 47 pp

Forestry Canada Fire Danger Group FCFDG (1992) Development and structure of the Canadian Forest Fire Behavior Prediction System. Information Report ST-X-3, Forestry Canada, Science and Sustainable Development Directorate, Minister of Supply and Services Ottawa, Ontario, 62 pp

Keane RE, Reinhardt ED, Scott J, Gray K, Reardon JJ (2005) Estimating forest canopy bulk density using six indirect methods. Can J For Res 35:724–739

Keane RE, Frescino TL, Reeves MC, Long J (2006) Mapping wildland fuels across large regions for the LANDFIRE prototype project. In: Rollins MG, Frame C (eds) The LANDFIRE prototype project: nationally consistent and locally relevant geospatial data for wildland fire management. USDA Forest Service Rocky Mountain Research Station, pp 367–396

Keane R, Gray K, Bacciu V, Leirfallom S (2012a) Spatial scaling of wildland fuels for six forest and rangeland ecosystems of the northern Rocky Mountains, USA. Landsc Ecol 27(8):1213–1234. doi:10.1007/s10980-012-9773-9

Keane RE, Gray K, Bacciu V (2012b) Spatial variability of wildland fuel characteristics in northern Rocky Mountain ecosystems. Research Paper RMRS-RP-98, USDA Forest Service Rocky Mountain Research Station, Fort Collins, Colorado, USA, 58 pp

Linn RR (1997) A transport model for prediction of wildfire behavior. Ph.D. dissertation, New Mexico State University, Las Cruces, New Mexico, USA

Nabel JEMS, Kirchner JW, Zurbriggen N, Kienast F, Lischke H (2014) Extrapolation methods for climate time series revisited—spatial correlations in climatic fluctuations influence simulated tree species abundance and migration. Ecol Complex. doi:http://dx.doi.org/10.1016/j.ecocom.2014.02.006

Nadkarni NM (1994) Diversity of species and interactions in the upper tree canopy of forest ecosystems. Am Zool 34(1):70–78. doi:10.1093/icb/34.1.70

Parsons RA (2006) Fuel 3D: a spatially explicit fractal fuel distribution model. In: Andrews PL, Butler BW (eds) Fuels management—how to measure success, Portland, OR USA. USDA Forest Service, Rocky Mountain Research Station, Proceedings RMRS-P-41. Fort Collins, CO, pp 51–66

Parsons RA, Mell WE, McCauley P (2010) Linking 3D spatial models of fuels and fire: effects of spatial heterogeneity on fire behavior. Ecol Model 222(3):679–691

Reeves MC, Kost JR, Ryan KC (2006) Fuels products of the LANDFIRE project. In: Andrews PL, Butler BW (eds) Fuels management—how to measure success, Portland OR. USDA Forest Service Rocky Mountain Research Station, Proceedings RMRS-P-41 Fort Collins, CO, pp 239–249

Reeves MC, Ryan KC, Rollins MC, Thompson TG (2009) Spatial fuel data products of the LANDFIRE project. Intl J Wildland Fire 18:250–267

Reinhardt E, Lutes D, Scott J (2006a) FuelCalc: a method for estimating fuel characteristics. In: Andrews PL, Butler BW (eds) Fuels management—how to measure success, Portland, OR. US Department of Agriculture, Forest Service, Rocky Mountain Research Station, Proceedings RMRS-P-41, Fort Collins, CO, pp 273–287

Reinhardt E, Scott J, Gray K, Keane R (2006b) Estimating canopy fuel characteristics in five conifer stands in the western United States using tree and stand measurements. Can J For Res-Rev Can Rech For 36(11):2803–2814. doi:10.1139/X06-157

Rothermel RC (1972) A mathematical model for predicting fire spread in wildland fuels. Utah Research Paper INT-115, United States Department of Agriculture, Forest Service, Intermountain Forest and Range Experiment Station, Ogden, 88 pp

Rothermel RC (1991) Predicting behavior and size of crown fires in the Northern Rocky Mountains. Research Paper INT-438, US Department of Agriculture, Forest Service Intermountain Forest and Range Experiment Station, Ogden, Utah, USA 46 pp

Rothermel RC (1993) Some fire behavior modeling concepts for fire management systems. In: Proceedings of the 12th Annual Conference on Fire and Forest Meteorology, Jekyll Island, GA, Society of American Foresters, Bethesda, MD USA, pp 164–171

Sando RW, Wick CH (1972) A method of evaluating crown fuels in forest stands. Research Paper NC-84, United States Department of Agriculture, Forest Service North Central Forest Experiment Station, Saint Paul, Minnesota, USA, 10 pp

Scott JH, Reinhardt ED (2001) Assessing crown fire potential by linking models of surface and crown fire behavior. Research Paper RMRS-RP-29, USDA Forest Service Rocky Mountain Research Station, Fort Collins, CO, 59 pp

Scott JH, Reinhardt ED (2002) Estimating canopy fuels in conifer forests. Fire Manage Today 62(4):45–50

Scott JH, Reinhardt ED (2005) Stereo photo guide for estimating canopy fuel characteristics in conifer stands. General Technical Report RMRS-GTR-145, USDA Forest Service Rocky Mountain Research Station, Fort Collins, CO, 47 pp

Stocks BJ, Alexander ME, Wotton BM, Stefner CN, Flannigan MD, Taylor SW, Lavoie N, Mason JA, Hartley GR, Maffey ME, Dalrymple GN, Blake TW, Cruz MG, Lanoville RA (2004) Crown fire behaviour in a northern jack pine-black spruce forest. Can J For Res 34(8):1548–1560. doi:10.1139/x04-054

van Wagner CE (1977) Conditions for the start and spread of crown fire. Can J For Res 7:23–34

van Wagner CEV (1993) Prediction of crown fire behavior in two stands of jack pine. Can J For Res 23:442–449

Ward DE (1995) Smoke emissions from biomass burning. In: Hassol SJ, Katzenberger J (eds) Elements of change. Changes in global vegetation patterns and their relationship to human activity. Aspen Global Change Institute, Aspen, pp 107–111

Waring RH, Running SW (1998) Forest ecosystems: analysis at multiple scales, 2nd edn. Academic, San Diego

Zhang T, Lichstein JW, Birdsey RA (2014) Spatial and temporal heterogeneity in the dynamics of eastern US forests: implications for developing broad-scale forest dynamics models. Ecol Model 279:89–99. doi:http://dx.doi.org/10.1016/j.ecolmodel.2014.02.011

Chapter 5
Fuel Moisture

Water is the driving force of all nature
Leonardo da Vinci

5.1 Background

Moisture content is the most dynamic feature of most wildland fuel types and it influences nearly all fire processes, especially ignition, combustion, and smoldering. The amount of moisture in the fuel is called the fuel moisture content (FMC) defined as the mass of water per unit mass of dry material and is often expressed as a percent. FMC is a major factor determining the heat produced by a wildland fire (see Table 2.2). High fuel moistures slow the rate of burning and fuel consumption in a number of ways. Heat from the fire must first volatilize the water bound in the fuel by boiling it off as gas before it can ignite the organic fuel (Simard 1968). This steals heat away from ignition and combustion processes, and if there is not enough heat to drive off a sufficient amount of moisture, then ignition will not occur. High moisture values may also cause increased particle thermal conductivity (how fast heat can pass through an object) and volumetric heat capacity (heat output per unit volume of fuel) that may further dampen ignition and combustion (Nelson 2001). Fuel moisture also reduces flame temperature thereby increasing smoldering combustion and the production of char while retarding the rate of consumption. Because moisture reduces both ignition potential and combustion temperatures, there is a corresponding increase in the time it takes a fuel particle to burn (residence time). Water vapor created from the heat of combustion may surround the fuel particle and dilute the available oxygen thereby also retarding combustion (Simard 1968). These complex interactions contribute to a lack of ignition, or if ignited, to lower heat emissions and longer burning times. This, in turn, can result in less flaming combustion, decreased fuel consumption, and increased smoldering. This is why every fire behavior model has fuel component FMCs as input parameters (Matthews 2013).

Three significant fuel characteristics affect moisture of live and dead fuel particles (Nelson 2001). The chemical composition of the fuel dictates its ability to attract and hold water molecules from the surrounding environment; a fuel property

R. E. Keane, *Wildland Fuel Fundamentals and Applications*,
DOI 10.1007/978-3-319-09015-3_5

called *hydroscopy*. In general, wood is mostly composed of cellulose (~50%), hemicellulose (~20%), and lignin (~20%); needles have less lignin and more cellulose (Chap. 2). First, cellulose has a greater ability to hold water than lignin because of its chemical structure (i.e., more hydroscopic). The ability of the fuel particle to hold water dictates the rate of moisture loss. Second, the internal cell structure of the fuel strongly influences moisture dynamics and it differs for live and dead fuels. Cell walls of most dead fuels are hygroscopic. Water molecules that are attracted to and eventually adhere to cell walls become bound water and have low vapor pressure (Schroeder and Buck 1970). Free water consists of those water molecules that are not bound to the chemical structure of the fuel. Third, the physical properties of the fuel ultimately control moisture retention. These include all of those properties described in Chap. 2, but especially particle density, size, and shape. Dense, large, round coarse fuels dry slower than less dense, small fuels that have high surface areas.

Live fuels have completely different moisture dynamics than dead fuels. In short, live fuel moistures are dictated by the ecophysiological processes of transpiration and soil water dynamics, while dead fuel moistures are driven by the physical process of evaporation. Both live and dead fuel moisture dynamics are driven by the gradient of vapor pressure (humidity) from the particle to the atmosphere; water molecules tend to migrate to drier conditions. Dead and live FMCs also have a complex spatial distribution because the biophysical factors that control FMC dynamics vary widely across space as they interact with plants, necromass, weather, and topography.

Although this chapter discusses fuel moisture in the context of how it is used to predict fire behavior and effects, live and dead fuel moisture dynamics interact with many other ecological and physical processes. Moist fuels, for example, decompose more rapidly and are more susceptible to fragmentation than dry fuels. Dry fuels intercept more precipitation and reduce water availability for plant growth. The primary objective of this chapter is to familiarize the reader with the biophysical processes that control fuel moisture, how they relate to wildland fuel dynamics (Chap. 6), and how to measure or estimate fuel moistures. A more comprehensive discussion of live and dead fuel moisture dynamics for fire managers is provided in Schroeder and Buck (1970) and a more scientific treatment is provided by Nelson (2001).

5.2 Dead Fuel Moisture Dynamics

Water moves through dead wood fuel particles via three mechanisms—capillarity forces, infiltration, and diffusion. *Capillary forces* draw water through fuel particles via fine capillaries in cell walls and cell structure. *Infiltration* involves the flow of free water through a fuel particle via gravitational forces. The primary mechanism is *diffusion* in which water in gaseous form (vapor) diffuses into and through a fuel particle driven by a moisture gradient; water vapor is drawn from areas of high moisture to areas of lower moisture to achieve equilibrium. Water vapor can either condense on cell walls or continue to diffuse through cell voids to any area of lower moisture. The steeper the particle water vapor gradient, the faster the diffusion process. Diffusion is governed by cell structure (i.e., pore space, cell wall composition), which varies by plant life-form and species of the dead particle, and particle age (stage of decomposition). In summary, water and water vapor move through a fuel particle in a progression of processes consisting of evaporation of water from cell walls, diffusion of the water vapor across cell voids and through cell walls, condensation on another cell wall, and bound water transport across a cell wall, all aided by capillary forces and infiltration. Eventually, water is evaporated from the particle surface into the atmosphere and lost from the fuel particle.

Two physical properties of the fuel represent these processes in modeling dead fuel moisture dynamics within a particle. Particle *permeability* is the ability of water to flow through cell cavities or how fast water can be transported across the particle. Cell structure, chemical composition, size, shape, and degree of decomposition all play a role in influencing the permeability of water through a dead fuel particle. Moisture *diffusivity* is the potential for the flow of water molecules across cell walls and is mostly governed by the degree of hygroscopy of the cell walls within the dead fuel particle.

To simplify complex physical FMC processes, dead fuel moisture dynamics is mostly governed by the water vapor pressure difference (dry to wet) between the atmosphere and the fuel particle. This difference is primarily driven by temperature, relative humidity, and the presence of water on the particle surface (Matthews 2013). If the atmosphere–particle vapor pressure difference is positive, such as when the temperature is high, relative humidity is low, and fuel is wet, then water is lost from the particle into the atmosphere through a process known as *desorption*. The free water in the intercellular space is easily evaporated to the atmosphere, but the vapor pressure deficit (difference in humidity of cell spaces and the air) must be great to drive bound water from the cell walls. *Adsorption* occurs when atmospheric humidity increases and dry fuel particles gain moisture as water molecules adhere to the particle surface resulting in increasing FMC (Simard 1968). The processes of desorption and adsorption introduce an important aspect of dead FMC dynamics—the FMC for any dead fuel particle is greatly dependent on its past conditions. This is called fuel moisture *hysteresis* in which today's fuel particle's FMC is dependent on past FMCs. As a result, most empirical equations that simulate fuel moisture include the previous days' FMC values (Viney 1991).

If there is rain, then free water on the particle surface can be *absorbed* by the particle until either the fuel particle has reached the fiber saturation point (fully saturated with water) or the precipitation stops and free water on the particle surface evaporates. Water absorption into fuel particles is a slow process, especially for large woody fuels, that depends on the many factors including initial moisture content, particle condition (percent rot, amount fragmentation), and type of particle (wood, grass, needle litter). As a result, the amount of rainfall may be unimportant because, during heavy rainfall events, most of the precipitation may run off or evaporate from the particle before being absorbed. Therefore, the duration of

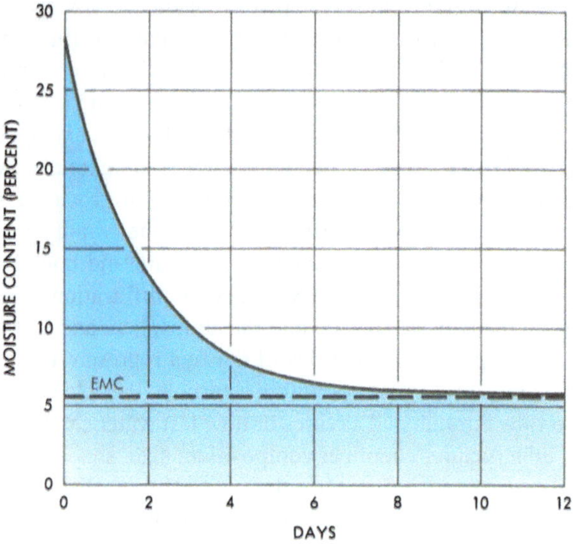

Fig. 5.1 Drying curve of a 5 cm-deep litter layer under temperature and humidity conditions for which the equilibrium moisture content (EMC) is 5.5 %. (Schroeder and Buck 1970)

rainfall, not the amount, is often used in fuel moisture modeling (Viney 1991). Dew can also condense on a fuel particle provided that the particle's temperature is at dew point and this condensed free water can be directly absorbed by the particle. Moisture from the ground can also be conducted to a fuel particle if the ground is wet, and water evaporated from the ground surface can condense on the cooler fuel particle (Pyne et al. 1996).

Any discussion of dead FMC dynamics requires the introduction of another term to illustrate how fuel dries over time. The equilibrium moisture content (EMC) is the final FMC of a dead fuel particle when it is exposed to constant temperature and humidity conditions for a period of time. When the difference in vapor pressure between air and the fuel particle is zero, the moisture of a dead fuel particle is in equilibrium with the constant environment (Fig. 5.1). EMC curves differ for each type and size of particle as well as for each set of temperature and humidity conditions. Most fuel particles rarely reach EMC because the surrounding environment changes rapidly and desorption rates are slow. The process of adsorption and desorption can also be illustrated using the relationship of EMC to relative humidity (Fig. 5.2). Low humidities allow desorption to decrease EMCs resulting in drying fuels (lower FMCs), while higher humidities increase adsorption to increase EMCs and ultimately FMCs. One application of EMC is in the construction of drying curves (Fig. 5.2) that show how the relative humidity affects the EMC at a constant temperature, in this case, of 27 °C (80 °F) of a fuel particle (needle) during both the adsorption (when the air is moister than the particle) and desorption processes.

Fire scientists have condensed the complex process of dead woody fuel moisture dynamics into the time-lag concept discussed briefly in Chap. 3. A *time lag* is the time it takes for a fuel particle to lose 63 % of the difference between its initial moisture content and its EMC under constant conditions (80 °F, 20 % relative humidity)

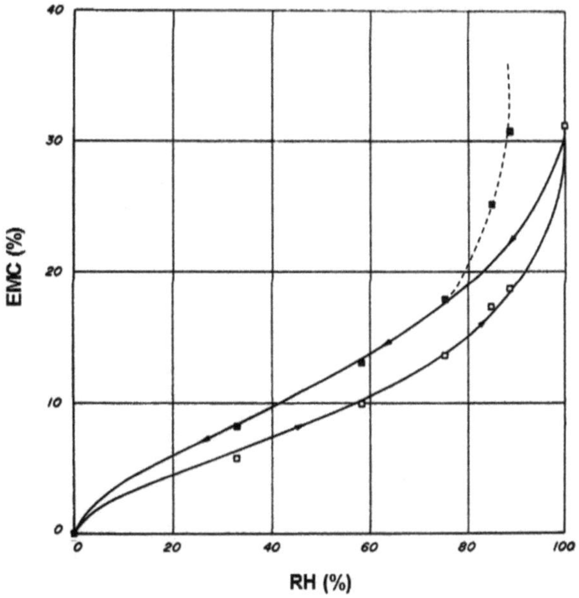

Fig. 5.2 The equilibrium moisture content (EMC) of sugar maple wood at a constant temperature of 27 °C (80 °F), but different levels of humidity when the humidity in the air is greater than in the dead fuel (adsorption) and when the air is drier than the fuel (desorption; Djolani 1970)

assuming an exponential drying function. One hour (1-h) woody fuel particles, for example, lose approximately two thirds of their moisture in 60 min. It takes about four time lags to get to 95 % of the EMC, historically set at 4.5 % (Fosberg et al. 1970). Time lags increase logarithmically with the diameter of the woody fuel particle; 10-inch diameter particles take 10 times longer to dry (100 days) than 1-inch particles (1 day; Fig. 5.3). Larger particles take longer to dry because diffusion has to pull water vapor from the wetter areas inside the wood to the drier, outside areas of the wood particle, which can only happen when the air is quite dry. As a result, the inside wood of larger particles (1000 h fuels) will often be wetter than the outside wood.

Fuel moisture dynamics are quite different for duff and litter as compared to wood. Time lags for litter and duff are highly variable and depend on depth, type, stage of decomposition, and a host of other biophysical factors. A 2 -inch-deep litter layer in a ponderosa pine stand, for example, might have a time lag of 48 h or the equivalent of a 1.4 in diameter woody fuel (Schroeder and Buck). Hydraulic properties of duff and litter mostly depend on the degree of decomposition with slightly decomposed material having large pore spaces resulting in increased hydraulic conductivity and faster drying and wetting (Plamondon et al. 1972). As decomposition proceeds, the size of the organic particles decreases resulting in smaller pores with higher moisture retention properties and slower drying. Hydraulic conductivity decreases significantly with the degree of decomposition. Duff bulk density is a useful indicator of pore size distribution and hence, can represent hydraulic conductivity and moisture retention in organic soils. Duff layers with high bulk densities, for example, may take longer to dry.

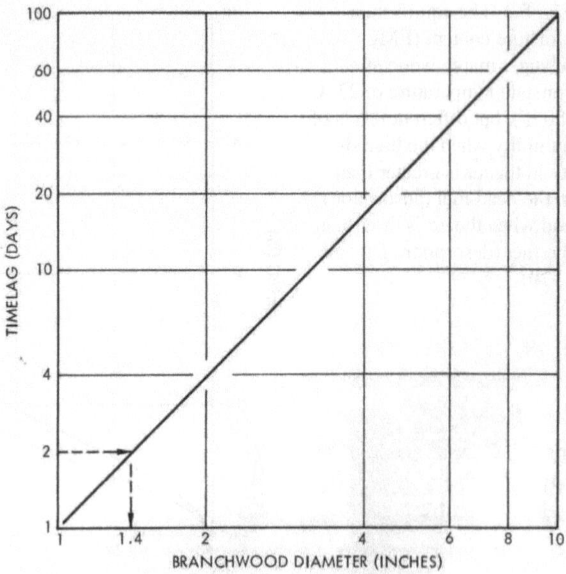

Fig. 5.3 The timelags of different diameter woody fuel particles. (Schroeder and Buck 1970)

5.3 Live Fuel Moisture Dynamics

Unlike dead fuel moisture, live fuel moistures are dictated by the responses of living plants to their surrounding environment (Pyne et al. 1996). Plants grow by transforming water, carbon dioxide, and the sun's energy into biomass via photosynthesis (Eq. 2.1) and they maintain this living biomass through the process of respiration, which also uses water. Plants get the water needed for respiration and photosynthesis from the soil though a series of complex processes known as water transport. When the ambient air is dry, there is a difference in vapor pressure (amount of water in the air) from the atmosphere to the stomatal cavities in the plants, photosynthetically active biomass (foliage). This difference creates a moisture gradient that pulls water from the foliar cells into the cell void and then through the stomata and into the atmosphere via a process known as transpiration (Campbell 1977). The movement of water out of the cell walls and into the intercellular spaces sets up a diffusion gradient and provides the tensional pressure to pull water through living cells from the roots, through the xylem, to the foliage, and out to the atmosphere.

Three factors control water transport in plants and therefore live fuel moisture dynamics: (1) osmotic forces caused by diffusion of water across the plant's cellular membranes, (2) capillary tension forces, and (3) diffusion across cell voids. Water enters the roots through diffusion when the transpirational pull, often called plant water potential, is greater than the soil water potential (forces that bind water to soil particles). Water moves through cells via osmosis and through cell voids by diffusion and capillary forces. The diffused water is then pulled up to the foliage through xylem conductive tissue in the wood by tensional pressure caused by plant water potential that is driven by the vapor pressure gradient from the air to the plant. As the soil water becomes depleted, the diffusional gradient must become greater to exert more tensional pressure to overcome the tensional forces that the soil exerts on the bound water. The moisture content of live fuels comes from the free and bound water and water vapor in plant's cellular void spaces, cell walls, and conducting tissue. This water is under tension due to transpirational pull and other forces such as capillary and surface tension. If soil water is abundant, plants open their stomata thereby initiating the vapor pressure gradient to pull soil water through cells for use in photosynthesis and respiration. However, there always needs to be some water within the plant to maintain respiration and keep the plant living, so live FMCs rarely get as low as dead FMCs. Most plants in fire environments have the ability to shut their stomata to control water loss that ensures the plant stays alive thereby keeping live FMCs high (Waring and Running 1998).

Many exogenous environmental factors control water transport in plants. As mentioned, humidity of the atmosphere is the primary engine that draws water from the soil and dictates the rates of transpiration (Campbell 1977). Soil properties, such as depth and percent rock, sand, silt, and clay will dictate soil water-holding capacity, permeability, and flow rates that then control the amount of free and bound water available for the plant throughout the year (Eagleson 1978). Precipitation is of critical importance in keeping the soil plenum full. Enough water has to fall so that it is not evaporated, intercepted by the live foliage, or absorbed by the ground fuels so that it can eventually seep into the soil and be available for plants. Increases in solar radiation may increase photosynthesis resulting in greater water usage and earlier water deficits contributing to lower FMCs. Topography is also important in that it will influence subsurface water flows to and away from the soil plenum, and it controls radiation, snowfall, and drainage dynamics.

While live fuel moisture dynamics are more complex than dead fuel moisture, live fuel moistures are also more stable over longer time periods. In general, live fuels, especially new plant growth, have the highest fuel moistures during the growing season when water is abundant and the new cells consist mostly of water with little structural material (Fig. 5.4). As cell walls harden and cell growth consists mostly of structural tissue, the mass of organic matter increases and the mass of water remains stable. As a result, the ratio of dry matter to water increases thereby lowering relative moisture content but not plant moisture. This is why moisture contents of older foliage are rarely as high as new foliage (Fig. 5.4). As soil water becomes scarce, water in the cells and cell voids becomes depleted causing major to minor decreases in live FMCs depending on the plant species. However, live foliar moisture rarely goes below the level at which plant cells would die from lack of moisture. Chrosciewicz (1986) found that moisture contents of jack pine and black spruce were approximately 120–130 % in the spring and fall, but dropped to 90 % in the summer. Live FMCs of actively growing foliage and branchwood can be as high as 300 % during the growing season when water is plentiful (Pyne et al. 1996). Conifers and many temperate shrubs may have live FMCs that reach a minimum of 90–100 % during the driest parts of the year. Some xeric shrubs, such as sagebrush, can have FMCs that are as low as 30–50 % during the dry season. The spatial pattern of live FMCs across an area

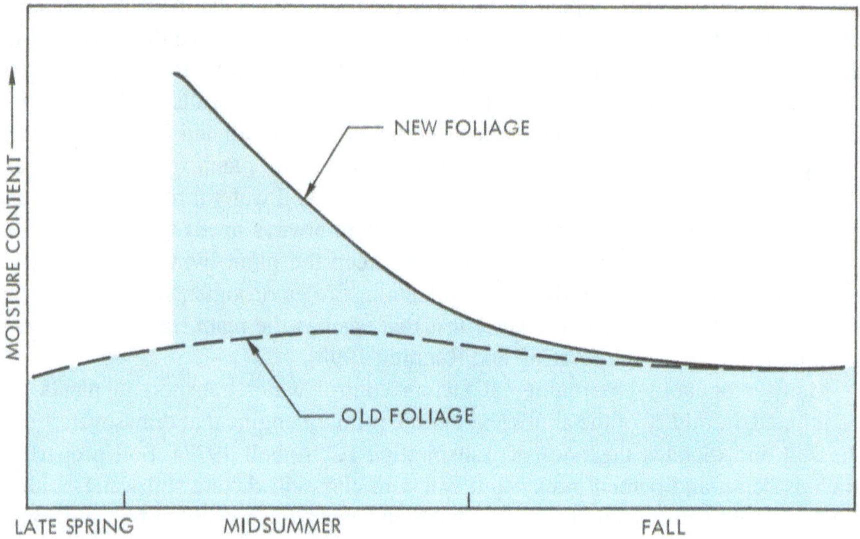

Fig. 5.4 The live fuel moisture content (FMC) of new and old pine needle foliage at different times of the year. Old foliage moisture content does not change throughout the year. (Schroeder and Buck 1970)

can be quite heterogeneous because most forest and range communities consist of a wide diversity of species in different stages of development and phenology (curing), each having different FMC dynamics.

Plant ecology is just as important as the environment when understanding live fuel moisture dynamics. Phenology is important because seasonal changes in plant development influence live FMC throughout the year. Live plant FMCs in the growing season, for example, are always much higher than FMCs of plants in the dormant season. Foliae age in conifers is also important; older needles have thicker, hardened cell walls (Fig. 5.4). There are also great differences among species in their ability to sustain foliage under low moisture conditions. Conifers, for example, have higher minimum foliar FMC values (>90 % FMC) in the fire season than xeric shrub species (>30 % FMC; Kauffman and Martin 1985). And FMCs can be quite different in a single plant; FMCs are typically higher in the upper third of a conifer's crown than the lower third, which is probably because of the abundance of new needles in the crown top (Brown 1978). Many perennial plants, such as grasses and forbs, allow their foliage to die when FMCs get too low through curing, while other species simply close their stomata to reduce transpiration and limiting water loss thereby maintaining FMC values.

The influence of live fuel FMC on combustion processes is still not well understood (Jolly 2007). Why do dead fuels cease burning when FMCs are above 30 %, while tree crowns burn when their moisture content is above 100 %? Several factors may contribute to this dilemma. First, water in live fuels is under tension thereby having a higher pressure and lower boiling point. Second, the chemical composition of live fuels is different than dead fuels and the many dissolved compounds in the bound and free water of live fuels may influence ignition (Jolly et al. 2012; McAllister et al. 2012). Bulk densities and packing ratios of canopy live fuels may facilitate rapid ignition processes to generate great heat outputs that overwhelm moisture effects. Hopefully, future research will provide answers to these important live fuel moisture questions.

5.4 Moisture of Extinction (M_x)

The moisture of extinction (M_x) is the dead or live fuel moisture at which combustion cannot continue (Chap. 2, 3). Trabaud (1976) defined M_x as the maximum FMC above which a fire cannot be sustained and found 45 % as threshold value; combustion did not occur above that value or it was delayed more than 15 min. In general, studies have found that fire spread usually stops when dead fuels have an FMC greater than 30 %, yet live fuels with FMCs above 100 % can support a spreading fire (Scott et al. 2014). Rothermel (1972) first formulated the effect of moisture content on the burning rate by defining a threshold M_x, above which fire cannot be sustained. The concept of extinction moisture is difficult to define in field and laboratory experiments because it varies by diverse fuel (size, shape, density) and environmental (temperature, humidity, wind) factors. As mentioned, FMC increases specific heat and thermal conductivity of fuel so that more heat is absorbed by the fuel particles surface layer to drive out moisture, delaying preheating and ignition of fuel until it reaches ignition temperature. If there is too much water, ignition will not occur. The most comprehensive M_x model was provided by Wilson (1985) based on laboratory data using milled wood sticks and shaved excelsior. Ideally, M_x should be an emergent property of fire models that simulate combustion using mechanistic physical process; simulated combustion should cease when moistures are too great. However, most fire models do not have the resolution and detail to accurately simulate the extinction of combustion due to high moistures. Instead, MEs are considered static parameters in combustion models and are assigned as properties of a fuel or particle (Chap. 2). Rothermel (1972), for example, used the ratio of FMC to M_x to calculate a dampening coefficient that reduced fire intensity (Table 2.2).

5.5 Measuring Moisture Content

The most common method of measuring FMC is called gravimetric sampling where fuel is dried in an oven. Fuel moisture is calculated from the difference in fuel weight before and after drying. In general, this method involves collecting fuels in the field and weighing them as soon as possible or placing them in a waterproof container and storing the container in a cold or cool place for transport to the laboratory. Once in the laboratory, fuels are first weighed and then dried in an oven set at temperatures that range from 50 to 105 °C until their weight is stable, typically requiring times that range from overnight to days depending on the fuel component. The dried fuel is then weighed and the difference between field-measured fuel weight and oven-dried weight is the mass of water lost in the drying process. This water loss is usually divided by the oven-dried fuel mass to get FMC in units of percent (Matthews 2013). The best results for this method occur when samples are dried at lower temperatures (80–85 °C) for longer time periods (48–72 h), which prevent unwanted changes to the organic structure of the fuel particle, especially live fuels and litter.

There are some problems with using the oven-dry method for estimating FMC (Simard 1968). First, fuel samples take a long time to process, making this method difficult to apply in real-time fire management operations, such as prescribed fires and wildfire suppression, when FMCs need to be quickly estimated. Second, care must be taken in both the field and the laboratory to ensure accurate fuel moisture measurements; collected samples must be immediately put into a plastic bag and put in a cool environment to prevent adsorption and water loss before weighing in the laboratory. Drying organic materials in a hot oven may also volatilize other chemical constituents of the fuel particles, especially live foliage, resulting in an unwanted loss of organic biomass and a slight overestimate of FMC. On the other hand, if the oven temperature is too low, some water may be retained in the cell structure and FMC estimates may be underestimated. Some woody fuel particles are so large that it may take days to completely dry the particles, while small fuels, such as twigs, take only hours to dry. These disparate drying times make obtaining a comprehensive set of FMCs for all fuel components difficult in an operational setting.

Several other field methods have been used to estimate FMCs in the field. Banks and Frayer (1966) correlated the angle at which a pinaster pine needle broke to the needle's FMC to create the "leaf-bending method" for quick and easy field use, albeit only for that forest type and only for needles. The "speedy moisture meter" used by Dexter and Williams (1976) measures the pressure of acetylene gas evolved from mixing finely divided fuels with calcium carbide to estimate FMC, but its use was limited because it required careful preparation of fuel particles (Marsden-Smedley and Catchpole 2001). Another FMC method is to pulse an electric charge through the fuel particle and measure electrical signals, such as resistance, impedance, or capacitance. FMC is then calculated from a calibration curve that relates the electric signal to FMC (Stamm 1927). The Wiltronics TH Fine Fuel Moisture Meter, for example, estimates FMC from the electrical resistance of a fine fuel sample (Chatto and Tolhurst 1997). Gravimetric sampling, however, is still the most common and reliable method to calculate FMC (Viney 1991; Chatto and Tolhurst 1997).

Because FMC is difficult to conveniently and accurately measure in the field, it is often estimated using indirect techniques. One indirect method is to estimate FMC for dead woody fuels using fuel analogues or moisture indicator sticks. A set of four wooden dowels of standard size and weight are placed at a weather station and weighed at various time intervals to approximate fuel moisture for 10-h woody fuel component. The current and past moistures of these dowels are related empirically to estimate FMCs for other fuel components (Schroeder and Buck 1970). Simulation modeling also provides an indirect way to estimate FMCs where computer programs are used to simulate fuel moisture dynamics from weather and fuel inputs (Nelson 2001). In fact, many fire managers use simulated FMCs from the National Fire Danger Rating System (NFDRS) models to estimate local FMCs for their fuels although the simulation of dead fuel FMCs is much better than live fuel FMCs (Anderson 1976). And, in recent years, several studies have found useful empirical relations between FMC and satellite-derived variables in several ecosystems (Paltridge and Mitchell 1990; Chladil and Nunez 1995; Chuvieco and Martin 1994). Dead fuel FMC estimation from remotely sensed data is complex for two reasons: (1) dead fuels are under the vegetation canopy and, therefore, cannot be directly sensed remotely, and (2) dead fuels do not show changes in green coloration from water variations and, consequently, are less sensitive to changes in reflectance. For this reason, FMCs for grasslands was more effectively estimated with higher precision than other fuels because FMC variations in grasslands have a greater influence on those variables that affect plant reflectance (such as chlorophyll content or leaf area index; Paltridge and Mitchell 1990; Hardy et al. 1999).

References

Anderson HE (1976) Fuels—the source of the matter. Paper presented at the Air quality and smoke from urban and forest fires. In: Proceedings of the international symposium, Colorado State University, Fort Collins, CO, 1976

Banks WG, Frayer HC (1966) Rate of forest fire spread and resistance to control in the fuel types of the eastern region. U.S. Department of Agriculture, Forest Service, Fire Control Notes, vol 27. Washington, DC

Brown JK (1978) Weight and density of crowns of rocky mountain conifers. United States Department of Agriculture, Forest Service Intermountain Forest and Range Experiment Station, Research Paper INT-197 Ogden, UT USA 56 pp

Campbell GS (1977) An introduction to environmental biophysics. Springer-Verlag, New York

Chatto K, Tolhurst KG (1997) Development and testing of the wiltronics TH fine fuel moisture meter. Fire Management Branch, Department of Natural Resources and the Environment Victoria, Australia ISBN: 0730666980

Chladil MA, Nunez M (1995) Assessing grassland moisture and biomass in Tasmania—the application of remote sensing and empirical models for a cloudy environment. Int J Wildland Fire 5(3):165–171

Chrosciewicz Z (1986) Foliar moisture content variations in four coniferous tree species of central Alberta. Can J For Res 16:157–162

Chuvieco E, Martin MP (1994) Global fire mapping and fire danger estimation using AVHRR images. Photogramm Eng Remote Sens 60(5):563–570

Dexter BD, Williams DF (1976) Direct field estimation of fine fuel moisture content. Aust For 39(2):140–144. doi:10.1080/00049158.1976.10675649

Djolani B (1970) Hysteresis and secondary phenomena in sorption of water in wood at temperatures of 5°, 21°, 35°, and 50°C. Notes Rech Dep Exploit Util Bois Univ Laval (8):58

Eagleson PS (1978) Climate, soil, and vegetation. Water Resour Res 14(5):705–776

Fosberg MA, Lancaster JW, Schroeder MJ (1970) Fuel moisture response—drying relationships under standard and field conditions. For Sci 16(1):121–128

Hardy CC, Burgan RE, Ottmar RD (1999) A database for spatial assessments of fire characteristics, fuel profiles, and PM10 emissions. J Sustain For 11(1–2):229–244

Jolly WM (2007) Sensitivity of a surface fire spread model and associated fire behaviour fuel models to changes in live fuel moisture. Int J Wildland Fire 16(4):503–509. doi:http://dx.doi.org/10.1071/WF06077

Jolly WM, Parsons RA, Hadlow AM, Cohn GM, McAllister SS, Popp JB, Hubbard RM, Negron JF (2012) Relationships between moisture, chemistry, and ignition of Pinus contorta needles during the early stages of mountain pine beetle attack. Forest Ecology and Management 269 (0):52–59. doi:http://dx.doi.org/10.1016/j.foreco.2011.12.022

Kauffman JB, Martin RE (1985) Shrub and hardwood response to prescribed burning with varying season, weather, and fuel moisture. In: Proceedings from the 8th conference on fire and forest meterology, Detroit, MI. Society of American foresters, Bethesda, MD, pp 279–286

Marsden-Smedley JB, Catchpole WR (2001) Fire modelling in Tasmanian buttongrass moorlands. III. Dead fuel moisture. International Journal of Wildland Fire 10(2):241–253. doi:http://dx.doi.org/10.1071/WF01025

Matthews S (2013) Dead fuel moisture research: 1991–2012. Int J Wildland Fire. 23(1):78–92. doi:http://dx.doi.org/10.1071/WF13005

McAllister S, Grenfell I, Hadlow A, Jolly WM, Finney M, Cohen J (2012) Piloted ignition of live forest fuels. Fire Safety J 51:133–142. doi:http://dx.doi.org/10.1016/j.firesaf.2012.04.001

Nelson RM Jr (2001) Water relations of forest fuels. In: Johnson EA, Miyanishi K (eds) Forest fires: behavior and ecological effects. Academic, San Diego, pp 79–149

Paltridge GW, Mitchell RM (1990) Atmospheric and viewing angle correction of vegetation indices and grassland fuel moisture content dervied from NOAA/AVHRR. Remote Sens Environ 31:121–135

Plamondon PA, Black TA, Goodell BC (1972) The role of hydrologic properties of forest floor in watershed hydrology. National symposium on Watersheds in Transition. American Water Resources Association, pp 341–348

Pyne SJ, Andrews PL, Laven RD (1996) Introduction to wildland fire, 2nd edn. Wiley, New York

Rothermel RC (1972) A mathematical model for predicting fire spread in wildland fuels. United States Department of Agriculture, Forest Service, Intermountain Forest and Range Experiment Station, Ogden, Utah Research Paper INT-115, 88 pp

Schroeder MJ, Buck CC (1970) Fire weather...A guide to application of meteorological information to forest fire control operations. U.S. Department of Agriculture, Forest Service, Agric Handb No. 360 Washington, DC 229 pp

Scott AC, Bowman DMJS, Bond WJ, Pyne SJ, Alexander ME (eds) (2014) Fire on earth: an introduction. Wiley, Chichester

Simard AJ (1968) The moisture content of forest fuels—I- A review of the basic concepts. Department of Forestry and Rural Development, information report FF-X-14 Ottawa, Ontario, Canada, 44 pp

Stamm AJ (1927) The electrical resistance of wood as a measure of its moisture content. Ind Eng Chem 19(9):1021–1025

Trabaud L (1976) Inflammabilid et Combustibilid des Principales Esptces des Ganigues de la Rdgion MMiterran6wnw. Oecol Plant 11(2):117–136

Viney NR (1991) A review of fine fuel moisture modelling. Int J Wildland Fire 1(4):215–234

Waring RH, Running SW (1998) Forest ecosystems: analysis at multiple scales, 2nd edn. Academic, San Diego

Wilson RA, Jr (1985) Observations of extinction and marginal burning states in free burning porous fuel beds. Combustion Science and Technology 44, 179–193.

Chapter 6
Ecology

The first law of ecology is that everything is related to everything else.
Barry Commoner American Scientist

6.1 Fuel Dynamics

The most notable feature of wildland fuel beds is that they are always changing in space and time because complex ecological processes are continuously interacting with the fuelbed. Live and dead biomass is constantly being added and removed by various ecological processes (Harmon et al. 1986). Spring snowstorms, for example, may break tree and shrub branches and double fuel loadings in a few hours, while decomposition may take decades to centuries to reduce coarse woody debris (CWD) loadings. The annual shed of leaves and thousands of small woody twigs from trees in forested stands creates significantly different fine woody debris (FWD) spatial distributions than the infrequent toppling of large tree boles to create logs or CWD fuel types. As a result, wildland fuel landscapes can be described as shifting mosaics of hierarchically intersecting fuel characteristics. This dynamic character of fuelbeds across space and time is perhaps the single most important fuel characteristic to understand in fire management today because it influences strategic fuel management considerations such as fuel treatment longevity and effectiveness, fire return intervals, and smoke potential. In fact, Anderson (1976) wrote, "recognizing fuel complexes as storehouses with irregular annual additions and withdrawals of energy provides a basis for fire and smoke management." There are numerous processes that control fuel dynamics, but this chapter will present four of the major processes that most influence spatial and temporal distributions of fuels: biomass production, deposition, decomposition, and disturbance (Fig. 6.1). In wildland fuel science, many have assumed that fuels are closely related to vegetation characteristics, but this only makes sense for the first two processes (production and deposition) and completely ignores the role that decomposition and disturbance play in fuel bed development.

R. E. Keane, *Wildland Fuel Fundamentals and Applications*,
DOI 10.1007/978-3-319-09015-3_6

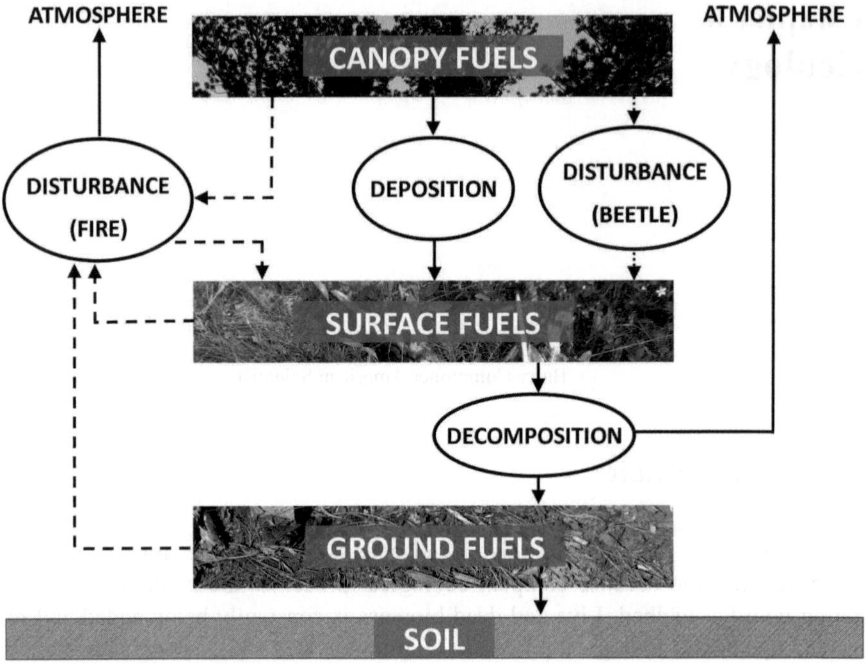

Fig. 6.1 Interaction of the four major processes controlling fuel dynamics on the surface and canopy fuel strata

6.1.1 Biomass Production

Wildland fuels accumulate as a result of the establishment, growth, and mortality of vegetation. The rate of biomass development, often called productivity (kg m^{-2} yr^{-1}), is dictated by the plant species available to occupy the area and the area's physical environment (e.g., climate, soils, and topography; Waring and Running 1998). The total amount of light energy transformed by photosynthesis into organic compounds is called gross primary productivity (GPP; Waring and Schlesinger 1985). The actual amount of biomass (ultimately, fuel) that is produced is called net primary production (NPP) and it is less than GPP because some energy is used for plant maintenance respiration (MR), which is the energy required to keep living biomass alive, and also for growth respiration (GR), which is the energy required to grow new biomass (NPP = GPP − MR − GR). The sum of MR and GR is called autotrophic respiration (AR = MR + GR). Live biomass can accumulate in many forms, such as roots, leaves, stems, and reproductive organs and propagules. The total amount of accumulated live biomass within an area is referred to as standing crop (SC, kg m^{-2}), with the live biomass that occurs aboveground called the aboveground standing biomass (ASB) and the biomass that occurs in the soil called belowground standing biomass (BSB). Over time, portions of ASB are shed or die and get deposited on the ground to become dead surface fuels or necromass.

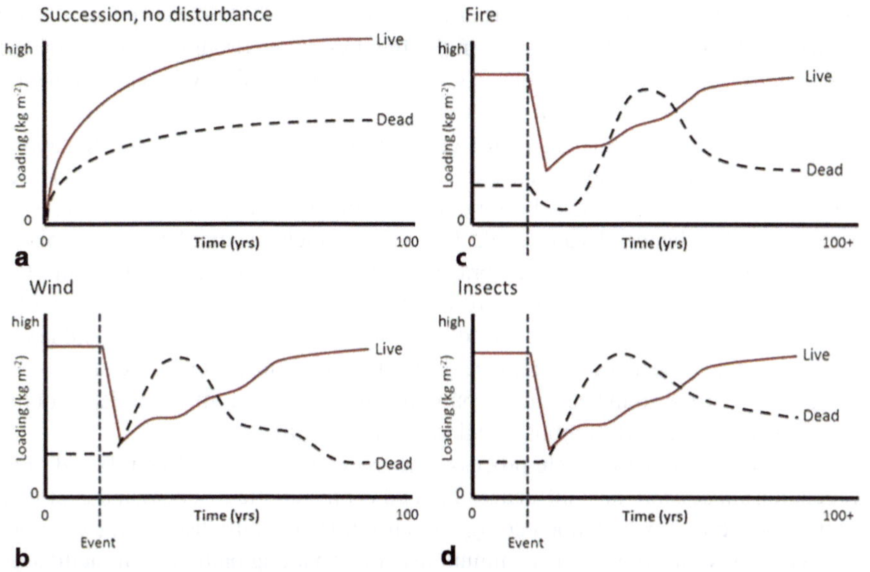

Fig. 6.2 The accumulation of live (*solid line*) and dead (*dashed line*) fuels in the absence of disturbance and after various disturbances. **a** only succession and vegetation development and no disturbance (note development starts with bare ground), **b** after a major wind event, **c** after a severe fire event, and **d** after a mountain pine beetle outbreak

Below- and aboveground necromass is eventually decomposed by microbes and soil macrofauna (see Sect. 6.1.3) to create duff and the expended energy during decomposition is called heterotrophic respiration (HR). The difference between total productivity (GPP) and the sum of both autotrophic and HR is called net ecosystem production (NEP = GPP − HR − AR or NEP = NPP − AR). The annual NEP for a fuelbed generally hovers around zero over long time periods, but in periods when NEP is positive, fuels are generally accumulating. When NEP becomes negative, it is likely the result of a disturbance or mortality event has probably reduced GPP because of the decrease in photosynthetic potential when plants are killed or damaged, and increased HR through the subsequent decomposition of the recently killed plant material. Negative NEP can persist for months or years depending on the severity of the disturbance (Diffenbaugh and Field 2013).

Wildland fuels accumulate in different ways depending if they are alive or dead (Fig. 6.2a). Live fuels accumulate through the completion of life-cycle processes of living organisms, primarily plants. Plants first become established on a site via reproduction and regeneration; they then grow as a result of photosynthesis and respiration, and they then die from a number of causes but most often disturbances, especially in fire-prone ecosystems (see Sect. 6.1.4). Live fuel biomass accrues as plants become established and grow bigger and taller, thereby increasing live fuel loading (increasing ASB) in both surface and canopy fuel layers. Increases in ASB from annual NPP eventually level off as photosynthetic gains are balanced by autotrophic respiration losses (Waring and Running 1998) and that balance point is dictated by a multitude of factors in the biophysical environment, primarily the interactions of plant characteristics (e.g., species, size, density) with environmental conditions (e.g., available water, nutrients, temperature, and sunlight; Fig. 6.2a). The fuelbed becomes taller as plants grow in height and expand into the canopy layer. And, as more plants become established, the canopy layer becomes denser (higher CBD). Thus, fuelbeds become heavier, higher, and denser with increasing time since disturbance (Keane et al. 2002). Dead fuels accumulate as live plants shed various parts, such as needles, branches, and fruits, or die from various mortality agents. As a result, dead fuel deposition rates will tend to increase with increasing live fuels, and these deposited dead fuels will tend to be more diverse and larger with increasing time since disturbance (Habeck 1985; Muller 2003). Dead fuels decompose over time so the accumulation will eventually level off when depositional gains are balanced by decompositional losses (Fig. 6.2a).

The interaction of life-cycle processes across plant species, often referred to as vegetation *succession,* also influences fuelbed characteristics. Succession is often a contentious term in vegetation ecology because it is generally used in the Clementsian context where vegetation communities develop along pathways of facilitation (i.e., one community paves the way for the next; Connell and Slayter 1977; Clements 1916). This book, however, uses the term to describe any form of vegetation or fuel development in any direction (progressive or retrogressive). In succession, species that have evolved disturbance adaptations, such as thick bark, sprouting, and deep roots, will tend to persist or populate disturbed sites by surviving the disturbance and/or dispersing propagules into the disturbed area from great distances (Noble and Slatyer 1980). These disturbance-adapted species are usually unable to grow in heavy shade (shade-intolerant) because they have developed adaptations to grow quickly and dominate disturbed areas and take advantage of the ample sunlight available in postburn environments. As time since disturbance increases, shade-tolerant species often become established and grow in shaded areas because they have higher growth rates in low sunlight than shade-intolerant species. These shade-tolerant species first dominate the understory and then eventually outcompete the shade-intolerant individuals in the overstory. There is a general increase in shade-tolerant living biomass (ASB) with succession and this living biomass becomes more homogeneously distributed across all canopy layers as competition-adapted, shade-tolerant species replace disturbance-adapted shade-intolerant species (Keane et al. 2002). Some shade-tolerant species have unique morphological characteristics that also influence fuelbed characteristics. Shade-tolerant species usually have greater leaf biomass to harvest more light under low light conditions (Bazzaz 1979) and this leaf biomass tends to be distributed across the entire length of the plant. Shade-tolerant conifers, for example, tend to have crowns that are denser and closer to the ground (Brown 1978). Branches on some shade-tolerant conifers tend to be shorter in length and smaller in diameter than those of shade-intolerant trees, especially those adapted to disturbance (Brown 1978).

Plant growth and succession cause great changes in fuelbed characteristics over time (Fig. 6.2a). For canopy fuels, CH, CBD and CFL will generally increase, and

CBH will decrease with advancing succession or time since disturbance because of increasing plant density, continued growth, and species compositional shifts (Reinhardt et al. 2006). It is more difficult to describe surface fuel layer changes because of the interaction of deposition with decomposition and climate, but in general, loading for all dead fuel components will tend to increase with succession and plant development until decomposition rates approximately match deposition rates with the greatest accumulations occurring on sites with the lowest decomposition rates (Keane 2008b). In general, the larger the fuel component (e.g., CWD), the longer it takes for that component to reach this deposition–decomposition equilibrium and also the greater the variability in fuel properties, especially particle density, because large fuels are often in a wider range of decay stages than smaller fuels as a result of their slow decomposition rates. Surface fuel depth and bulk density also increase with succession as understories become dense with shrubs, herbs, and trees and dead fuels accumulate. Conversely, the low light conditions created by some dense shade-tolerant overstories in late successional stages may also deter shrub, herbaceous, tree regeneration resulting in lower surface fuel depth in some forested ecosystems. However, fuelbeds are rarely created solely from the processes of growth and succession, but instead are formed by the complex interactions of vegetation productivity with deposition, decomposition, disturbance and the physical environment (Collins and Roller 2013).

6.1.2 Deposition

Deposition is defined as the release of live and dead aerial biomass to fall to the ground to become dead surface fuels (Fig. 6.1). Many ecological studies refer to this process as *litterfall,* which is confusing because, in this book, litter is a term for a specific fuel component. Fuel deposition can result from normal plant ecophysiological and phenological processes, such as leaf shed and turnover, and also from the effects of endemic and exogenous disturbances (see Sect. 6.1.4). Light winds, for example, may dislodge senescing biomass and cause it to fall to the ground close to the plant, while strong winds may detach both dead and live biomass and transport it great distances. The interactions of fuel particle characteristics (e.g., size, shape, density) with wind distributions (e.g., speed, direction, duration) create unique fuel component patterns across the landscape (Keane et al. 2012a).

Rates of deposition (kg m^{-2} yr^{-1}) differ greatly by fuel component and ecosystem type. Most studies estimated only foliage or log deposition rates because they are the easiest to measure and comprise the majority of deposited necromass (Harmon et al. 1986; Vogt et al. 1986; Table 6.1). FWD additions to the forest floor are rarely reported even though they may be the most important to fuels management and fire behavior prediction because they influence fire spread. Example deposition rates are presented in Table 6.1 for US Rocky Mountain ecosystems and other ecosystems of the world. What is striking about these numbers is the great amount of organic material that falls to the forest floor each year. Douglas-fir stands, for

Table 6.1 Deposition and decomposition rates for material measured in various western US ecosystems (from Keane 2008a). Fuel components are coarse woody debris (CWD); twigs and branches are assumed to be less than 8 cm (3 inches) in diameter

Ecosystem	Fuel component	Deposition rate (kg m^{-2} yr^{-1})	Decay constant k (yr^{-1})	Canadian province or US state	References
Pinus ponderosa	Logs	0.03	0.05	AZ	Avery et al. 1976; Klemmedson 1992
	Foliage	0.29	0.05, 0.14, 0.08–0.18	CA, AZ	Bray and Gorham 1964; Yavitt and Fahey 1982; Stohlgren 1988; Klemmedson et al. 1990; Hart et al. 1992
Pseudotsuga menziesii	Logs	0.70, 0.45, 0.04, 0.15–0.45, 0.28	0.006–0.050	OR, WA	Wright and Lauterback 1958; Grier and Logan 1977; Gottfried 1978; Sollins 1982; Harmon et al. 1986; Spies et al. 1988; Edmonds and Eglitis 1989; Harmon and Hua 1991
	Twigs, branches	–	0.007–0.129, 0.06–0.14, 0.06, 0.005–0.05	WA	Fogel and Cromack 1977; Edmonds et al. 1986; Edmonds 1987; Edmonds and Eglitis 1989; Christiansen and Pickford 1991; Maguire 1994
	Foliage	0.50, 0.17–0.33, 0.114–0.177	0.005–0.010, 0.44, 0.27, 0.41–0.56, 0.178–0.284	OR, BC, WA	Dimock 1958; Turner and Long 1975; Fogel and Cromack 1977; Edmonds 1979, 1991; Graham 1982; Means et al. 1985; Sollins et al. 1987; Harmon and Hua 1991; Trofymow 1991; Prescott et al. 2000
Pinus contorta	Logs	0.02	0.027, 0.082, 0.0016–0.0027, 0.115, 0.015	CO, AL	Alexander 1954; Pearson 1987; Taylor et al. 1991; Busse 1994; Laiho and Prescott 1999; Kueppers et al. 2004
	Twigs	–	0.055	AL	Taylor et al. 1991; Prescott et al. 1993
	Foliage	0.362	0.115, 0.14, 0.09–0.11	AL, WY	Yavitt and Fahey 1982; Taylor et al. 1991; Berg and Ekbohm 1993; Stump and Binkley 1993; Laiho and Prescott 1999
Tsuga heterophylla	Logs		0.016–0.018	OR	Graham 1982
	Twigs, branches		0.08–0.24	WA	Edmonds 1987
	Foliage		0.3–0.5	BC	Keenan et al. 1996
Abies lasiocarpa and *Picea* spp.	Logs		0.001–0.0015	CO	Kueppers et al. 2004
	Foliage	0.2–0.23	0.09–0.17	AL	Taylor et al. 1991; Laiho and Prescott 1999; Prescott et al. 2003

example, appear to deposit at least around 0.4 kg m^{-2} yr^{-1} (~0.2 kg m^{-2} yr^{-1} for foliage and 0.2 kg m^{-2} yr^{-1} for logs), which is about 3 % of the total aboveground biomass (Table 4.1). Keane (2008b) found that foliage deposits on productive sites were often 10–60 % of the total litter loading. Deposition rates for logs (Table 6.1) are usually measured from historical tree mortality and snag fall rates over time, which assumes tree fall is the only input to log accumulation. Large branches and treetops, however, also contribute to log inputs to the forest floor in some ecosystems (Harmon et al. 1986). While values in Table 6.1 provide estimates of fuel deposition rates, actual deposition rates are highly dependent on local vegetation conditions, such as species composition, stand structure, disturbance history, and biophysical environment (Keane 2008a).

Each fuel component has a different spatial and temporal pattern of deposition on the ground. The finest fuels, such as foliage and small twigs, tend to be more evenly distributed over time and space (Keane 2008b). This is because the foliage and small supporting branches are constantly shed by plants and their small size facilitates long distance and homogeneous dispersal by wind and gravity. Coarse fuels, however, have higher variability in deposition rates from year to year and from place to place; coarser fuels tend to fall infrequently, usually a result of extreme events, such as a fires, high wind, or heavy snow load. Because of their size, coarse fuels fall close to parent plants creating patchy patterns that often reflect the patterns of the live vegetation. Many years may pass before a large branch or tree bole falls to the ground and these fallen larger particles tend to be more scattered across the landscape in heterogeneous patterns.

6.1.3 Decomposition

Decomposition is the process whereby dead organic biomass on the soil surface is broken down into smaller particles and simpler forms (Millar 1974; Swift et al. 1979; Fig. 6.1). There are three main sequential mechanisms of decomposition (Marra and Edmonds 1996). Decomposition usually begins with *leaching,* where soluble carbon compounds are dissolved in water (precipitation) and this solution eventually seeps into the soil. *Fragmentation* then physically splits and breaks fallen plant material into smaller pieces, creating greater surface areas for microbial colonization and consumption. Fragmentation is typically accomplished by invertebrate fauna in the soil such as nematodes, insects, mites, and earthworms, and also by abiotic weathering (freezing and thawing, drying and wetting). Insects, especially termites, ants, bark beetles, and wood borers, play important roles in the fragmentation of large wood particles and they also introduce fungal decomposers to fragmented material (Harmon et al. 1986). Microbes, primarily bacteria and fungi, then invade the disintegrated plant matter, termed *detritus,* composed mainly of lignin and microbial byproducts. Microbial *respiration* further alters the chemical structure of the detritus and continues decomposition. Respiration is an oxidation process; much like fire (see Chap. 2), in that it transforms carbon compounds and oxygen into carbon dioxide, water, and energy to sustain metabolic microbial processes (see Eq. 2.2).

Fuels are composed of a wide variety of organic compounds that decompose at different rates. Microbes decompose simple sugars to CO_2 and water relatively quickly and completely, but the decomposition of complicated organic compounds in forest ecosystems is rarely complete (Meentemeyer 1978). Lignin, for example, has a complex chemical structure that slows the rate of microbial breakdown because it can only be decomposed by certain fungi. This recalcitrant organic matter accumulates on and in the soil as humus to become the duff fuel component (Chap. 3). Every dead fuel component is in various stages of decay, but once material from a fuel particle has been broken down by leaching, fragmentation, and respiration to contain mostly lignin, it is mostly unrecognizable as a fuel particle and usually lands in the duff for the last stages of advanced decay, mainly from microbial respiration.

Decomposition alters fuel particle properties in a number of ways. Decay of woody fuel usually decreases particle densities (specific gravities), decreases heat content, and increases surface area to volume ratios (SAVR) as cracks develop (Harmon et al. 2008). As decay advances and decomposed organic material moves to the duff, the mineral content and moisture of extinction also tend to increase. Duff layers often have high mineral contents because of two factors: (1) the mixing of mineral soil into the duff by soil macrofauna and (2) the release of minerals into the duff via decomposition (Keane 2008b).

Decomposition rates are usually calculated using two approaches. In the first approach, a parameter k is used to describe rates of decomposition using the following formula from Olson (1963):

$$\frac{L_t}{L_o} = e^{-kt}, \tag{6.1}$$

where L_t is the loading at time t, L_0 is loading at the start or when $t = 0$, and t is time (yr). The exponential function represents the effect of the remaining material (L_t) having a greater content of recalcitrant lignin, making decomposition difficult and longer. The second approach uses a mass loss rate (MLR, %yr^{-1}), calculated as the difference in loading over a unit time period ($MLR = 100*(L_0 - L_t)/L_0$). MLR must be calculated over long time periods to ensure the estimates include both the rapid decay of solubles, hemicellulose, and cellulose, and the prolonged decay of lignin. As with deposition, studies of decomposition rates are mostly for foliage and large log material, especially in the western USA (Table 6.1; MLR can be approximated in Table 6.1 for the first 10 years of decomposition from the k value by solving Eq. 6.1 using $t = 10$ and $L_0 = 10$).

6.1.4 Disturbance

Disturbance dictates fuel dynamics in most ecosystems and disturbance history is often considered the most important factor for predicting and describing fuel characteristics (Brown and Bevins 1986; Keane et al. 2012a). Disturbance is defined here as an event, force, or agent, of biological or nonbiological origin, that causes mortality to organisms and changes patterns in ecosystems (Pickett and White 1985). Disturbances can be endogenous, originating within the area of interest, or exogenous, coming from outside the area. Many types of disturbances influence fuel dynamics and each type has its own unique impacts on fuelbed characteristics. Disturbances may occur over short periods of time, such as wildland fire, or over longer periods of years or decades, such as insect and disease outbreaks. A disturbance event may impact an entire stand, such as when a crown fire kills all trees or shrubs, or it may affect only certain species or individuals, such as when mountain pine beetles kill specific pine tree host species within a stand. In addition, disturbances may kill entire plants or simply damage parts of them. Disturbances can also interact with each other, such as beetles and fire (Jenkins et al. 2008), or be totally independent of each other, such as blister rust and beetles (Six and Adams 2007).

A *disturbance regime* is a general term that describes the temporal and spatial characteristics of disturbances. Disturbance regimes are fundamentally different from individual disturbance events because regimes incorporate the cumulative effects of multiple disturbances over time. Thus, fuel dynamics are governed more by the disturbance regimes rather than any particular disturbance event. Disturbance regimes can be generally described by 11 characteristics (Table 6.2; Keane 2013). The disturbance *agent* is the entity that causes the disturbance, such as wind, fire, or beetles. Often, disturbance agents have a source that triggers the agent. Lightning can be a source for wildland fire and heavy snow loads may be the source for avalanches. Disturbances occur at a particular *frequency* that is often described over a period of time depending on scale and objective. Point-level measures, such as disturbance return interval and occurrence probability, describe the number of disturbance events experienced over time at one point on the landscape. Disturbance *intensity* is the level of the disturbance agent as it occurs on the landscape. Insect and disease intensities are often described by population levels; windthrow intensity can be described by wind speed. *Severity* is different from intensity in that it reflects the impact of that disturbance on the biophysical environment and it is an important attribute because it directly links to land management. The *sizes* (area) and *patterns* (spatial variability) of disturbance severity and intensity often govern landscape and fuelbed heterogeneity, which influences a wide variety of landscape characteristics such as connectivity, fragmentation, and diversity (Turner et al. 1997; Knight 1987). Pattern refers to the size, shape, and spatial distribution of disturbed patches. *Seasonality* reflects the annual time of occurrence typical of a disturbance event, important because it can influence plant and animal phenology, mortality, and post-disturbance recovery. Disturbance patterns are often influenced by the *duration* of the disturbance agent on the landscape, with durations ranging from seconds (wind)

Table 6.2 The eleven terms often used to describe disturbance regimes

Disturbance characteristic	Description	Example
Agent	Factor causing the disturbance	Mountain pine beetle is the agent that kills trees
Source, cause	Origin of the agent	Lighting is a source for wildland fire
Frequency	How often the disturbance occurs or its return time	Years since last fire or beetle outbreak (scale dependent)
Intensity	A description of the magnitude of the disturbance agent	Mountain pine beetle population levels; wildland fire heat output
Severity	The level of impact of the disturbance on the environment	Percent mountain pine beetle tree mortality; fuel consumption in wildland fires
Size	Spatial extent of the disturbance	Mountain pine beetles can kill trees in small patches or across entire landscapes
Pattern	Patch size distribution of disturbance effects; spatial heterogeneity of disturbance effects	Fire can burn large regions but weather and fuels can influence fire intensity and therefore the patchwork of tree mortality
Seasonality	Time of year of that disturbance occurs	Species phenology can influence wildland fires effects; spring burns can be more damaging to growing plants than fall burns on dormant plants
Duration	Length of time of that disturbances occur	Mountain pine beetle outbreaks usually last for 3–8 years; fires can burn for a day or for an entire summer
Interactions	Disturbances interact with each other, climate, vegetation, and other landscape characteristics	Mountain pine beetles can create fuel complexes that facilitate or exclude wildland fire
Variability	The spatial and temporal variability of the above factors	Highly variable weather and mountain pine beetle mortality can cause highly variable burn conditions resulting in patchy burns of small to large sizes

and minutes (avalanches) to days (fire) and years (insects). The *variability* of disturbance characteristics such as severity, frequency, and size, coupled with their *interaction* with existing patterns, duration, and seasonality of disturbance, as well as climate and vegetation, govern landscape dynamics and ultimately controls wildland fuels (Schoennagel et al. 2004).

The following sections provide simplistic descriptions of the effects of four common western North American disturbance regimes on fuelbed properties. It is important to note that these disturbances also interact with climate, vegetation, and other disturbances at the landscape level to uniquely modify fuel dynamics and create locally unique fuelbeds (Keane et al. 2014).

6.1.4.1 Wind

Wind influences fuel deposition in most ecosystems because of its high frequency, variable intensity, and disparate severities (Mitchell 2013). Many fuel particles would eventually fall to the ground without wind, but because wind is such a frequent phenomena, its action often dislodges most plant material long before it would fall on its own and then transports the detached fuel particles various distances. Wind effects range from chronic to acute. Trees respond to chronic wind exposure by thickening stems and structural roots, and by reducing shoot length thereby resulting in shorter and thicker branches and trunks that ultimately make unique surface fuel particles. Moderate acute wind may cause minor acute effects by freeing dead or dying plant material and depositing it over small areas. Strong winds, however, may strip live foliage and branch material from upright plants and deposit them great distances from the source plant (Schoennagel et al. 2012). Severe wind may also cause plant breakage or uprooting, a disturbance process known as "windthrow" or "blowdown," and these major effects create significantly different fuelbeds (Mitchell 2013). Hurricanes and typhoons may have winds that are so strong that they can topple most trees thereby moving canopy fuels directly to the ground surface over a short time period (Busing et al. 2009; Turner et al. 1997). The severity and frequency of windthrow depends on many site factors including soils, plant conditions (health, size), stand conditions (densities, size distribution), topography, management history, and, of course, wind characteristics (e.g., speed, duration, and direction).

Windthrow disturbances generally decrease canopy fuels, increase surface fuel loadings, and change surface fuelbed properties (Fig. 6.2b). Wind rarely reduces surface fuel loading as with some other disturbances. As a result, wind serves several ecological roles in fuel deposition. First, winds affect spatial distributions of fuels by blowing fuel particles either evenly or unevenly across an area, depending on particle size, wind speed and direction, and vegetation condition (species, size, morphology, health). Second, strong winds may collapse canopy fuels into the surface fuel layer thereby decreasing CBH, CBD, and CC but increasing surface fuel component loadings, bulk density, depth, and fuel component diversity (Vihnanek et al. 2009). Schoennagel et al. (2012) mention that hurricane winds were important historical drivers of fire regimes in the southern USA because they created continuous surface fuelbeds with heavy loadings that facilitated the spread of fire. And last, wind can create unique patterns of fuelbed properties across the landscape because wind events and their intensities vary spatially and temporally in forests from large-scale catastrophic episodes, such as hurricanes and typhoons, to landscape level occurrences, such as thunderstorms and tornadoes, to fine-scale perturbations, such as downdrafts and microbursts (Ulanova 2000). Each of these wind events can cause a patchwork of effects from minor branch breakage to minor windthrow damage to complete forest blowdown across landscapes.

6.1.4.2 Fire

Fire is one of the few disturbances with effects that can work both ways; fire can both reduce and increase fuel loadings depending on the characteristics of the burn, conditions of the prefire fuelbed and vegetation, and time since burn. Fire reduces live and dead fuel loadings by consuming them during combustion and smoldering phases (Chap. 3). The amount of fuel consumed by the fire depends on many factors, such as the bulk density and depth of the fuelbed along with the moistures, mineral contents, densities, and SAVRs of the various particles in that fuelbed (Albini and Reinhardt 1995; Call and Albini 1997; Chap. 2). Commonly, wildland fires consume nearly all of the fine fuels (litter, twigs, herbs, shrubs) and partially consume large branches, CWD, and duff (Call and Albini 1997). Fire often leaves a complex mosaic of residual fuel loadings because of microsite differences in fuel size, moisture, loadings, and weather (wind, temperature, radiation). Most forested ecosystems also experience decreases in CBD and increases in CBH after fire. Crown fires, for example, significantly reduce canopy fuels by direct consumption of all crowns, whereas surface and mixed severity fires may reduce CBD fuels by either killing those plants that are most susceptible to fire because they were small or maladapted to survive fire or by directly scorching plant parts. Microsite differences in surface and canopy fuel consumption become greater when fire intensities are low, but as deep drought, high temperatures, and strong winds increase fire intensity and spread, microscale consumption and plant mortality often become more homogeneous (Brown and Reinhardt 1991). Keane and Parsons (2010), for example, observed that unburned patches after low-intensity prescribed burns were mostly found in the shade of overstory trees where radiation was reduced so that fuels did not dry sufficiently to carry the fire.

Fire can also increase surface fuel loadings by killing or damaging plants but not consuming their biomass. The scorched, dead biomass eventually falls on the ground over time to increase surface fuel loadings and decrease canopy fuels (Martin et al. 1979; Fig. 6.2c). Surface fuel loadings may also increase as fires burn or smolder at the base of live trees and snags to consume enough stemwood so that the standing boles fall. Increases in surface fuel component loadings after fire have been

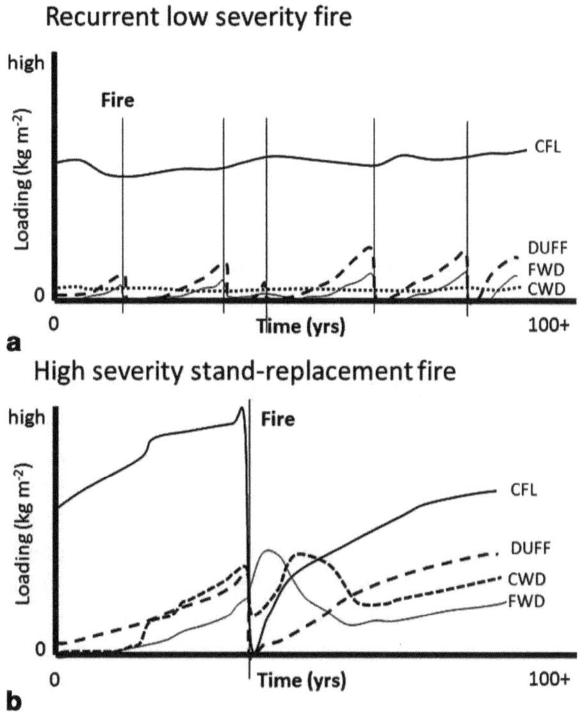

Fig. 6.3 Possible changes in fuelbed loadings for four fuel components and two wildland fire regimes: **a** frequent low severity fires, **b** stand-replacement fires. The fuel components are canopy fuel load (CFL: *thick line*), duff and litter (DUFF: *dashed line*), fine woody debris (FWD: *thin line*), and coarse woody debris (CWD: *dotted line*)

documented for hemlock/Douglas-fir forests (Agee and Huff 1987), southeastern US pine savannas (Thaxton and Platt 2006), and whitebark pine ecosystems (Keane and Parsons 2010).

Different fuel accumulation trajectories often result as a consequence of the distribution of severity, intensity, and pattern of fires in a fire regime (Fig. 6.3). For ecosystems with frequent fire regimes, repeated fires are constantly reduce fuels on the forest floor thereby keeping all dead fuel loads low (Fig. 6.3a). However, shrub and herbaceous loadings may become the dominant surface fuel components because they are adapted to survive fire and all other component loadings are low due to consumption by fire. After stand-replacement fires in forests, the dead snags of the fire-killed trees eventually fall and create high log loads for many years or until the next fire (Fig. 6.3b). Fine woody fuels peak right after the fire, and then decline until established trees become large enough to contribute significant FWD. Shrub and herb biomass peaks in the open conditions left after the fire and declines as tree density and overhead shade increase. Litter and duff loadings slowly increase over time; litter increases as a result of increased growing biomass and duff increases as deposition is greater than decomposition over time.

6.1.4.3 Insects and Disease

Fuel dynamics after insect and disease outbreaks will vary with the agent and the intensity and severity of the outbreak (Table 6.1), but in general, insects and diseases kill or damage their hosts thereby reducing live canopy fuels and increasing surface fuels (Augusiak et al. 2013; Parker et al. 2006). Dead and damaged canopy material eventually falls to the ground and becomes part of the surface fuelbed, as has been shown after spruce budworm infestations (Azuma 2010), blister rust infections (Campbell and Antos 2000), gypsy moth invasions (Collalti et al. 2014), and bark beetle outbreaks (Parker et al. 2006). Unlike wind and fire, insects and pathogens usually target specific plant hosts and sizes for infection, and as a result, the distribution and abundance of host plants on the landscape becomes a major factor in subsequent fuel dynamics. This section will focus on the bark beetles and fungal pathogens as examples of two significant disturbances because it would be nearly impossible to cover the wide variety of insects and disease species that influence forest and rangeland fuel complexes throughout the world.

The insect species most extensively studied for their influence on fuel dynamics are bark beetles, and more specifically, the mountain pine beetle (Jenkins et al. 2008). Surface fuels after major mountain pine beetle outbreaks appear to change in a number of ways (Fig. 6.2d). First, there is a flush of fine fuels, primarily needles and small twigs, occurring in the few years following mortality (Hoffman et al. 2012). These decompose relatively quickly as the large branches and eventually the beetle-killed snags fall over the next several decades (Page and Jenkins 2007). Klutsch et al. (2014) found that FWD and CWD increased by 10% for every 1.0 m^2 ha^{-1} of basal area loss from endemic beetle mortality. Dead trees eventually topple as decomposition rots the root structure thereby increasing 100-h and 1000-h fuels.

Canopy fuel characteristics also change after a beetle outbreak. Foliage on fatally attacked pine trees first turns red then falls to the ground over the next 2–5 years. These red needles may have lower moistures and higher flammability than green foliage, possibly creating a higher potential for crown fires (Schoennagel et al. 2012; Jolly et al. 2012). But when these red needles fall leaving only dead pine tree bole and branches, which is often called the gray stage (4–10 years after an outbreak), the canopy characteristics of CBD, CBH, and CC decrease to reduce fire hazard, but the magnitude of the reduction in canopy fuels depends on the severity of the beetle outbreak (Hicke et al. 2012). Meanwhile, the space vacated by the beetle-killed trees may be quickly occupied by opportunistic herb, shrub, and most importantly, seedlings and saplings of shade-tolerant tree species to increase live fuel biomass and change canopy fuel characteristics (Kovacic et al. 1985). Many factors control the timing and magnitude of the live and dead fuel dynamics after beetle outbreaks. First, the level of tree mortality (severity) and tree species composition (pre-outbreak conditions) of the host stand dictates the potential amount of biomass that can be actually deposited on the ground (Hicke et al. 2012). Mixed species stands where 50% of the trees are pine host species and all host species were killed deposit less fuel than stands composed entirely of host species that experienced 100% mortality. Second, the biophysical setting and host species characteristics often dictate the timing and duration of litterfall and its subsequent decomposition. Beetle-killed ponderosa pine stands on dry sites, for example, may have slower deposition and higher snag fall rates than high elevation, cold lodgepole pine stands (Hoffman et al. 2012). And last, interactions with other disturbances, such as wind and fire, alter temporal and spatial patterns of beetle-killed fuel accumulation (Parker et al. 2006).

Few studies have documented fuel dynamics after pathogen epidemics but the net effect is similar to that for insects—there is an increase in surface litter and woody fuels and a corresponding decrease in canopy fuel characteristics as plants and plant parts are killed by pathogens (Parker et al. 2006). Metz et al. (2010) found increases in surface fuel loads after sudden oak death in coastal California oak forests, but these increases did not increase fire severity when site was burn in a wildfire. Root and stem rots were major factors that increased surface fuel loadings in ponderosa pine stands in the Black Hills of South Dakota, USA (Lundquist 2007). In redwood forests of California, Metz et al. (2013) found that increases in disease possibly caused by climate change have increased fuel loadings to foster fires that may not be survived by resident redwood trees. Fungal infections may alter more than fuel component loadings; the dead and dying trees may be composed of wood in advanced stages of rot so when they fall, the logs they create will probably have lower particle densities and higher SAVRs (Hartley 1958).

6.1.4.4 Grazing

Grazing, as it influences fuelbeds, involves the consumption of live and sometimes dead herbaceous, shrub, and tree fuels by native ungulates and domestic livestock. Grazing directly changes the surface fuel complex by reducing loadings of the fine herbaceous and shrubby fuel components, and possibly tree regeneration foliage (Zimmermann and Neuenschwander 1984) and indirectly changes surface and canopy fuels by facilitating increased tree regeneration due to reduced grass, shrub, and herb competition that in turn increases litter and woody material on the ground and biomass in the canopy (Miller and Rose 1999). Grazing effects on wildland fuels depend on many factors including the grazing species (agent), abundance, timing, and duration of grazing (intensity, seasonality), responses of the plant species being grazed (severity), and the climate (Asner et al. 2004). Grazing will tend to make fuels discontinuous and less connected, resulting in a more heterogeneous fuelbed pattern (Turner and Bratton 1987) unless areas are overgrazed.

Changes to fuelbed properties caused by grazing are unique to specific ecosystems and landscapes. Bachelet et al. (2000) and Briggs et al. (2002) found that woody species encroachment into grasslands was enhanced by grazing and held in check by fires, and that fires in encroached areas were often more intense and severe thereby killing more trees. However, Moreira et al. (2001) found that agricultural abandonment and the reduction of grazing increased shrubs and trees in Portugal. Grazing may also facilitate the invasion of exotic plants into the ecosystems thereby altering fuelbed characteristics and potential fire behavior (Keeley et al. 2003). Areas might be grazed so intensively that little surface fuel remains to ignite a fire so that they become fuel breaks thereby impeding the spread of fire, but as less palatable woody plants become established in heavily grazed areas, canopy and surface FWD may increase, and fire spread may eventually be possible, and the subsequent effects may be more severe than a grassland fire.

6.1.4.5 Disturbance Interactions

While effects of individual disturbances may be important to fuelbed characteristics in some ecosystems, it is really the interaction of disturbances that most influences fuel dynamics (Schoennagel et al. 2004; Bigler et al. 2005). These interactions may be direct and immediate, such as trees surviving mixed severity fire becoming more susceptible to subsequent windthrow because of their increased wind exposure, or they may be indirect and long term, such as mountain pine beetle mortality in pine forests allowing shade-tolerant species in the understory to increase CBD and lower the canopy layer (CBH), thereby creating the potential for more severe and intense crown fires. Interactions can be simple, such as where grazing reduces flashy fuels and thus limits fire spread and decreases fire intensity (Bachelet et al. 2000), or they can be quite complex, such as increases in woody plant density and size as a result of fire exclusion, which then increases transpiration to reduce water availability making the ecosystem more susceptible to tree mortality from insects and pathogens in heavy droughts that then fosters more intense fires (Bigler et al. 2005). The origin, condition, and future dynamics of most fuelbeds can only be understood by addressing the multiple interactions of disturbances with the biophysical environment.

There are many interesting examples of multiple disturbance interactions that influence fuel dynamics (Keane et al. 2014). Bark beetles and the fungus *Phaeolus schweinitzii* may interact with fire to alter lodgepole pine fuel dynamics in Oregon, USA (Geiszler et al. 1980). Windthrown trees may facilitate an increase in *Ips* spp. beetle populations that then kill surrounding live trees resulting in even heavier loadings and deeper fuelbeds (Goheen and Hansen 1993). Trees that remain after fuel reduction treatments may be more prone to wind damage that results in canopy fuel decreases and surface fuel increases (Reinhardt et al. 2008). Matson and Bart (2013) noted that root disease of spruce caused by *Inonotus tomentosus* may contribute to increased mortality from the spruce beetle (*Dendroctonus rufipennis*) in boreal and subboreal spruce forests of British Columbia but the interaction is highly governed by the condition of the vegetation. Bachelet et al. (2000) found that the interaction of fire with grazing maintained the pine savanna in the Black Hills of South Dakota USA. In southwestern US ponderosa pine forests, dwarf mistletoe (*Arceuthobium vaginatum*) weakens pine trees making them more susceptible to mountain pine beetle attack thereby increasing tree mortality and surface fuels that will then tend to foster more intense wildfires that will kill even more pines (Parker et al. 2006).

6.2 Landscape Ecology

The landscape ecology of wildland fuels is the interaction of the above processes across multiple space and time scales to create shifting mosaics of fuel conditions. Understanding the spatial and temporal distributions of fuels may provide a better understanding of the impact of various wildland fuel management activities on fuel properties and it also might help explain unexpected fire behaviors and effects (Parsons et al. 2010). It can also aid in developing effective fuel applications that integrate spatial variability in their design such as new fuel classifications (Chap. 7), sampling methods (Chap. 8), and geospatial data (Chap. 9). Patterns of fuel characteristics will be important inputs to the fire effects and behavior models of the future (Parsons et al. 2010; King et al. 2008).

While many have extensively studied the landscape ecology of fire (McKenzie et al. 2011), few have looked at the spatial and temporal relationships of the wildland biomass that fuels these fires. Reich et al. (2004) evaluated the spatial variability of several fuel components over a large landscape in the US Black Hills and found that the variability was correlated to topography and vegetation. Hiers et al. (2009) measured small-scale variations in surface fuel using LiDAR and found that fuelbed depths become spatially independent after small distances (0.5 m^2). Spatial variability of grasslands have been described in the context of population dynamics and restoration potential but have not been related to fuel characteristics (Peters et al. 2006). Theobald (2013) found that while fine-scale variation in fuels dictated fire behavior, the distribution of CWD dictated germination in longleaf pine ecosystems. While some studies have described fuel distributions across landscapes (Ferrari 1999; Jin 2004), few have actually quantified the variability of fuel properties across space (Jia et al. 2006; King et al. 2008; Miller and Urban 2000). And, while many have identified fuel continuity as a major spatial characteristic of wildland fuels (Knapp and Keeley 2006; Jenkins et al. 2012), few studies have evaluated fuel patterns using landscape metrics.

Several landscape metrics are important in describing fuel patterns. Contagion is the probability that a pixel, patch, or polygon will be adjacent to a pixel, patch, or polygon of the same attributes, such as the same fuel complex, while dispersion is the inverse of contagion. Patch density is the number of patches (area of homogeneous fuel characteristics) per unit area that indirectly represents patch size. Landscape shape indexes measure the irregularity of patch shapes using perimeter to area ratios. Few studies have described landscape pattern and patch dynamics of wildland fuels. Sturtevant et al. (2004) used proportion of the landscape, mean patch size, nearest neighbor distance, and juxtaposition as metrics to describe connectivity of high-risk fuel types in Wisconsin, USA. A major limitation of most landscape metrics is that they were designed for mapped categorical variables, such as categories in a fuel classification (Chap. 8), but variables to describe fuel properties are continuous variables and there are a limited set of landscape analyses to describe spatial distributions of continuous fuel properties and landscape metrics for continuous variables are more complex and difficult to interpret.

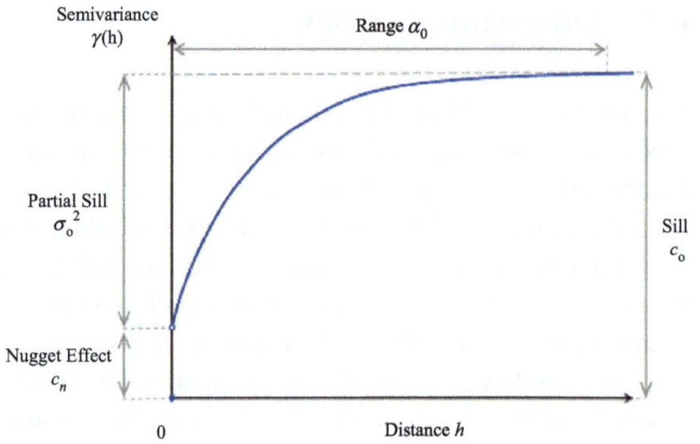

Fig. 6.4 The spatiovariogram and its characteristics from the SAS/STAT(R) 9.3 Users Guide. The nugget, sill, and range are commonly used to describe the spatial variability of an ecological characteristic

Spatial variability is an important landscape characteristic for describing landscape structure in continuous wildland fuel variables. It is often described using semivariograms; a descriptive technique that graphically represents the spatial continuity and autocorrelation of a spatial data set (Bellehumeur and Legendre 1998; Townsend and Fuhlendorf 2010). Semivariogram *range,* the distance where the variance curve first flattens (Fig. 6.4), is important in landscape ecology because it represents the spatial scale at which the entity of concern is best described in space, often called the inherent patch size (Fortin 1999). Using semivariograms, Keane et al. (2012a) estimated the spatial scale of individual fuel components on several US Rocky Mountain landscapes. They found that the smaller the fuel component, the finer the scale of spatial distribution. FWD varied at scales of 1–5 m, depending on the fuel particle, but CWD varied at 50–150 m and canopy fuel characteristics varied at 100–400m scales. This limited study shows that each fuel component has its own inherent scale and that this scale varies by biophysical environment, vegetation structure and composition, and time since disturbance. The implications of these findings are found in nearly all chapters of this book. Fuel classification effectiveness can be compromised because the variability of loadings across the unique spatial scales overwhelms the ability of the classification to uniquely identify disparate fuel classes (Chap. 7). Fuel sampling must account for the diverse scales of distribution between fuel components in sampling designs (Chap. 8), and fuel mapping must match the scale of mapping approaches and imagery to the scale of the fuel components being mapped to create accurate and consistent fuels layers (Chap. 9) (Fig. 6.4).

Another finding of the Keane et al. (2012b) study was the high spatial variability of a number of fuel properties within a site. The variability in loading for any fuel component was often twice the mean, even within a small homogeneous sampling area, and most other fuel properties, such as particle density, bulk density, and mineral content, also exhibited high variabilities (Table 3.2). They also found that this variability was not normally distributed but instead highly skewed towards the lower fuel values. Other findings were that this high variability could not be explained by any vegetation-based measurement or fuel loading estimate from any other fuel component; standard silvicultural measurements, such as basal area, tree density, and DBH, were not correlated to fuel component loadings, and none of the eight surface fuel components were correlated with each other. These findings provide valuable insight into why it is so difficult to create fuel applications and products that accurately predict fuel loadings—the high variability within a fuel component coupled with the fact that each component loading is independent of other component loadings and the spatial distribution of that variability is different for each fuel component often overwhelms statistical analyses (Keane et al. 2013).

The spatial variability of wildland fuel components over time directly impacts the fire regime, which in turn, has major ramifications for fire management. Landscape patches that have insufficient fuels to sustain fire spread, such as recently treated or burned patches, form fuel breaks that limit fire growth, reduce fire intensity, and minimize fire severity (Agee and Skinner 2005). The extent and spatial distribution of these burned patches on the landscape modify growth of future fires. This self-organizational property of wildland fire will be incredibly important in predicting future fire dynamics under climate change (McKenzie and Kennedy 2011; McKenzie et al. 2014). Fire frequencies, for example, may increase under warming climates only to a point when postburn patches limit fire growth. Eventually, dynamics of the fuel mosaic interact with fire to create landscapes that are self-organized and exhibit a unique fire regime (McKenzie et al. 2011). Fire and fuel management can use this ecological theory to develop management plans that effectively integrate wildfires, controlled wildfires, prescribed fires, and fuel treatments to minimize firefighting costs and maximize ecosystem resilience while still protecting homes and people (Reinhardt et al. 2008).

References

Agee JK, Huff MH (1987) Fuel succession in a western hemlock/Douglas-fir forest. Can J For Res 17(7):697–704

Agee JK, Skinner CN (2005) Basic principles of forest fuel reduction treatments. Forest Ecol Manag 211:83–96

Albini FA, Reinhardt ED (1995) Modeling ignition and burning rate of large woody natural fuels. Int J Wildland Fire 5(2):81–91

Alexander RR (1954) A comparison of growth and morality following cutting in old-growth mountain spruce stands. USDA Forest Service Rocky Mountain Forest and Range Experiment Station. Research Note RN-11. Fort Collins, CO USA. 11 pp

Anderson HE (1976) Fuels—the source of the matter. Paper presented at the Air quality and smoke from urban and forest fires—Proceedings of the international symposium, Colorado State University, Fort Collins, CO, 1976

Asner GP, Elmore AJ, Olander LP, Martin RE, Harris AT (2004) Grazing systems, ecosystem responses, and global change. Annu Rev Environ Res 29(1):261–299. doi:10.1146/annurev.energy.29.062403.102142

Augusiak J, Van den Brink PJ, Grimm V (2013) Merging validation and evaluation of ecological models to 'evaludation': a review of terminology and a practical approach. Ecol Model. doi:http://dx.doi.org/10.1016/j.ecolmodel.2013.11.009

Avery CC, Larson FR, Schubert HG (1976) Fifty-year records of virgin stand development in southwestern ponderosa pine. USDA Forest Service Rocky Mountain Forest and Range Experiment Station General Technical Report, Fort Collins, CO USA. 22 pp

Azuma D (2010) The effects of a western spruce budworm outbreak on the dead wood component in relation to ownership in forests of eastern Oregon. West J Appl For 25(4):176–180

Bachelet D, Lenihan JM, Daly C, Neilson RP (2000) Interactions between fire, grazing and climate change at Wind Cave National Park, SD. Ecol Model 134:229–244

Bazzaz FA (1979) The physiological ecology of plant succession. Annu Rev Ecol Syst 10:351–371

Bellehumeur C, Legendre P (1998) Multiscale sources of variation in ecological variables: modeling spatial dispersion, elaborating sampling designs. Landsc Ecol 13:15–25

Berg B, Ekbohm G (1993) Decomposing needle litter in *Pinus contorta* (Lodgepole pine) and *Pinus sylvestris* (Scots pine) monocultural systems—is there a maximum mass loss. Scand J For Res 8:457–465

Bigler C, Kulakowski D, Veblen TT (2005) Multiple disturbance interactions and drought influence fire severity in rocky mountain subalpine forests. Ecology 86(11):3018–3029

Bray JR, Gorham E (1964) Litter production in forests of the world. Advances in Ecological Research. Wiley and Sons, New York, pp 101–157

Briggs JM, Knapp AK, Brock BL (2002) Expansion of woody plants in tallgrass prairie: a fifteen-year study of fire and fire-grazing interactions. Am Midl Nat 147(2):287–294. doi:10.1674/0003-0031(2002)147[0287:eowpit]2.0.co;2

Brown JK (1978) Weight and density of crowns of rocky mountain conifers. United States Department of Agriculture, Forest Service Intermountain Forest and Range Experiment Station, Research Paper INT-197 Ogden, UT USA, 56 pp

Brown JK, Bevins CD (1986) Surface fuel loadings and predicted fire behavior for vegetation types in the northern rocky mountains. U.S. Department of Agriculture, Forest Service, Intermountain Forest and Range Experiment Station, Research Paper INT-358 Ogden, UT USA, 9 pp

Brown JK, Reinhardt E (1991) Estimating and regulating fuel consumption to manage smoke in the interior west. In: 11th conference on fire and forest meterology, Missoula, Montana. Society of American Foresters, Bethesda, MD USA, pp 419–429

Busing RT, White RD, Harmon ME, White PS (2009) Hurricane disturbance in a temperate deciduous forest: patch dynamics, tree mortality, and coarse woody detritus. Plant Ecol 201(1):351–363. doi:10.1007/s11258-008-9520-0

Busse MD (1994) Downed bole-wood decomposition in lodgepole pine forests of central Oregon. Soil Sci Soc Am J 58:221–227

Call PT, Albini FA (1997) Aerial and surface fuel consumption in crown fires. Int J Wildland Fire 7(3):259–264

Campbell EM, Antos JA (2000) Distribution and severity of white pine blister rust and mountain pine beetle on whitebark pine in British Columbia. Can J Bot 30:1051–1059

Christiansen EC, Pickford SG (1991) Natural abatement of fire hazard in Douglas-fir blowdown and thinning fuelbeds. Northwest Sci 65:141–147

Clements F (1916) Plant succession: an analysis of the development of vegetation, vol Publ. 242. Carnegie Institution of Washington, Washington, DC

Collalti A, Perugini L, Santini M, Chiti T, Nolè A, Matteucci G, Valentini R (2014) A process-based model to simulate growth in forests with complex structure: evaluation and use of 3D-CMCC forest ecosystem model in a deciduous forest in Central Italy. Ecol Model 272:362–378. doi:http://dx.doi.org/10.1016/j.ecolmodel.2013.09.016

Collins B, Roller G (2013) Early forest dynamics in stand-replacing fire patches in the northern Sierra Nevada, (California, USA). Landsc Ecol 28(9):1801–1813. doi:10.1007/s10980-013-9923-8

Connell JH, Slayter RO (1977) Mechanisms of succession and their role in community stabilization and organization. Am Nat 111:1119–1144

Diffenbaugh NS, Field CB (2013) Changes in ecologically critical terrestrial climate conditions. Science 341(6145):486–492. doi:10.1126/science.1237123

Dimock EJ (1958) Litter fall in a young stand of Douglas-fir. Northwest Sci 32:19–29

Edmonds RL (1979) Decomposition and nutrient release in Douglas-fir needle litter in relation to stand development. Can J For Res 9:132–138

Edmonds RL (1987) Decomposition rates and nutrient dynamics in small-diameter woody litter in four forest ecosystems in Washington, USA. Can J For Res 17:499–509

Edmonds RL (1991) Organic matter decomposition in western United States forests. In: Management and productivity of western-montane forest soils. USDA Forest Service Intermountain Research Station General Technical Report INT-280. Boise, ID, pp 118–125

Edmonds RL, Eglitis A (1989) The role of the Douglas-fir beetle and wood borers in the decomposition and nutrient release from Douglas-fir logs. Can J For Res 19:853–859

Edmonds RL, Vogt D, Sandberg D, Driver CH (1986) Decomposition of Douglas-fir and red alder wood in clear-cuttings. Can J For Res 16:822–831

Ferrari JB (1999) Fine-scale patterns of leaf litterfall and nitrogen cycling in an old-growth forest. Can J For Res 29:291–302

Fogel R, Cromack K (1977) Effect of habitat and substrate quality on Douglas-fir litter decomposition in western Oregon. Can J Bot 55:1632–1640

Fortin M-J (1999) Spatial statistics in landscape ecology. In: Klopatek JM, Gardner RH (eds) Landscape ecological analysis: issues and applications. Springer-Verlag, New York, pp 253–279

Geiszler DR, Gara RI, Driver CH, Gallucci VF, Martin RE (1980) Fire, fungi, and beetle influences on a lodgepole pine ecosystem of south-central Oregon. Oecologia 46:239–243

Goheen DJ, Hansen EM (1993) Effects of pathogens and bark beetles on forests. In: Schowalter T, Filip GM (eds) Beetle-Pathogen interactions in conifer forests. Academic, San Diego, pp 175–196

Gottfried GL (1978) Five-year growth and development in a virgin Arizona mixed conifer stand. USDA Forest Service Rocky Mountain Forest and Range Experiment Station, Research Paper RM-203. Fort Collins, CO USA

Graham RL (1982) Biomass dynamics of dead Douglas-fir and western hemlock boles in mid-elevation forests of the Cascade Range. Ph.D. Dissertation. Oregon State University, Corvallis, Oregon, USA. 121 pp

Grier CC, Logan RS (1977) Old-growth *Pseudotsuga menziesii* communities of a western Oregon watershed: Biomass distribution and production budgets. Ecol Monogr 47:373–400

Habeck JR (1985) Impact of fire suppression on forest succession and fuel accumulations in long fire interval wilderness habitat types. Proceedings:symposium and workshop on wilderness fire. U.S. department of agriculture, forest service, intermountain forest and range experiment station, Ogden, UT., USA, Missoula, MT, pp 76–85

Harmon ME, Franklin JF, Swanson FJ, Sollins P, Gregory SV, Lattin JD, Anderson NH, Cline SP, Aumen NG, Sedell JR, Lienkaemper GW, Cromack K, Cummins KW (1986) Ecology of coarse woody debris in temperate ecosystems. Adv Ecol Res 15:133–302

Harmon ME, Hua C (1991) Coarse woody debris dynamics in two old-growth ecosystems. Bioscience 41:604–610

Harmon ME, Woodall CW, Fasth B, Sexton J (2008) Woody detritus density and density reduction factors for tree species in the United States: a synthesis. USDA forest service northern research station, General Technical Report NRS-29 Newton Square, PA USA, 84 pp

Hart SC, Firestone MK, Paul EA (1992) Decomposition and nutrient dynamics of ponderosa pine needles in a Mediterranean-type climate. Can J For Res 22:306–314

Hartley CCP (1958) Evaluation of wood decay in experimental work. Collections

Hicke JA, Johnson MC, Hayes JL, Preisler HK (2012) Effects of bark beetle-caused tree mortality on wildfire. For Ecol Manag 271:81–90. doi:http://dx.doi.org/10.1016/j.foreco.2012.02.005

Hiers JK, O'Brien JJ, Mitchell RJ, Grego JM, Loudermilk EL (2009) The wildland fuel cell concept: an approach to characterize fine-scale variation in fuels and fire in frequently burned longleaf pine forests. Int J Wildland Fire 18(3):315–325. doi:10.1071/WF08084

Hoffman CM, Hull Sieg C, McMillin JD, Fulé PZ (2012) Fuel loadings 5 years after a bark beetle outbreak in south-western USA ponderosa pine forests. Int J Wildland Fire 21(3):306–312. doi:http://dx.doi.org/10.1071/WF11019

Jenkins MA, Hebertson E, Page W, Jorgensen CA (2008) Bark beetles, fuels, fire and implications for forest management in the Intermountain West. For Ecol Manag 254:16–34

Jenkins MJ, Page WG, Hebertson EG, Alexander ME (2012) Fuels and fire behavior dynamics in bark beetle-attacked forests in Western North America and implications for fire management. For Ecol Manag 275:23–34. doi:http://dx.doi.org/10.1016/j.foreco.2012.02.036

Jia GJ, Burke IC, Goetz AFH, Kaufmann MR, Kindel BC (2006) Assessing spatial patterns of forest fuels using AVIRIS data. Remote Sens Environ 102:318–327

Jin S (2004) Computer classification of four major components of surface fuel in Northeast China by image: the first step towards describing spatial heterogeneity of surface fuels by imags. For Ecol Manag 203:395–406

Jolly WM, Parsons RA, Hadlow AM, Cohn GM, McAllister SS, Popp JB, Hubbard RM, Negron JF (2012) Relationships between moisture, chemistry, and ignition of *Pinus contorta* needles during the early stages of mountain pine beetle attack. For Ecol Manag 269:52–59. doi:http://dx.doi.org/10.1016/j.foreco.2011.12.022

Keane RE (2008a) Biophysical controls on surface fuel litterfall and decomposition in the northern rocky mountains, USA. Can J For Res 38(6):1431–1445. doi:10.1139/X08-003

Keane RE (2008b) Surface fuel litterfall and decomposition in the northern rocky mountains, USA. USDA forest service rocky mountain research station, Research Paper RMRS-RP-70 Fort Collins, Colorado, 22 pp

Keane RE (2013) Disturbance regimes and the historical range of variation in terrestrial ecosystems. In: Editor-in-Chief: Simon AL (ed) Encyclopedia of biodiversity, 2nd edn. Academic Waltham, pp 568–581. doi:http://dx.doi.org/10.1016/B978-0-12-384719-5.00389-0

Keane RE, Parsons RA (2010) A management guide to ecosystem restoration treatments: whitebark pine forests of the northern rocky mountains. USDA forest service rocky mountain research station, General technical report RMRS-GTR-232 Fort Collins, CO. 123 pp

Keane RE, Veblen T, Ryan KC, Logan J, Allen C, Hawkes B (2002) The cascading effects of fire exclusion in the rocky mountains. In: Baron J (ed) Rocky mountain futures: an ecological perspective. Island Press, Washington, pp 133–153

Keane R, Gray K, Bacciu V, Leirfallom S (2012a) Spatial scaling of wildland fuels for six forest and rangeland ecosystems of the northern Rocky Mountains, USA. Landscape Ecol 27(8):1213–1234. doi:10.1007/s10980-012-9773-9

Keane RE, Gray K, Bacciu V (2012b) Spatial variability of wildland fuel characteristics in northern rocky mountain ecosystems. USDA forest service rocky mountain research station, Research Paper RMRS-RP-98 Fort Collins, Colorado USA, 58 pp

Keane RE, Herynk JM, Toney C, Urbanski SP, Lutes DC, Ottmar RD (2013) Evaluating the performance and mapping of three fuel classification systems using forest inventory and analysis surface fuel measurements. For Ecol Manag 305:248–263. doi:http://dx.doi.org/10.1016/j.foreco.2013.06.001

Keane RE, Miller C, Smithwick E, McKenzie D, Falk D, Kellogg L (2014) Representing climate, disturbance, vegetation interactions in landscape simulation models. Landscape Ecol (in press)

Keeley JE, Lubin D, Fotheringham CJ (2003) Fire and grazing impacts on plant diversity and alien plant invasions in the sourthern Sierra Nevada. Ecol Appl 13(5):1355–1374. doi:10.1890/02-5002

Keenan RJ, Prescott CE, Kimmins JP, Pastor J, Dewey B (1996) Litter decomposition in western red cedar and western hemlock forests on northern Vancouver Island, British Columbia. Can J Bot 74:1626–1634

Klemmedson JO (1992) Decomposition and nutrient release from mixtures of Gambel oak and ponderosa pine leaf litter. For Ecol Manag 47:349–361

Klemmedson JO, Meier CE, Campbell RE (1990) Litter fall transfers of dry matter and nutrients in ponderosa pine stands. Can J For Res 20:1105–1115

King KJ, Bradstock RA, Cary GJ, Chapman J, Marsden-Smedley JB (2008) The relative importance of fine-scale fuel mosaics on reducing fire risk in south-west Tasmania, Australia. Int J Wildland Fire 17(3):421–430. doi:10.1071/WF07052

Klutsch JG, Beam RD, Jacobi WR, Negrón JF (2014) Bark beetles and dwarf mistletoe interact to alter downed woody material, canopy structure, and stand characteristics in northern

Colorado ponderosa pine. For Ecol Manag 315:63–71. doi:http://dx.doi.org/10.1016/j.foreco.2013.12.024
Knapp EE, Keeley JE (2006) Heterogeneity in fire severity within early season and late season prescribed burns in a mixed-conifer forest. Int J Wildland Fire 15(1):37–45. doi:10.1071/Wf04068
Knight DH (1987) Parasites, lightning, and the vegetation mosaic in wilderness landscapes. In: Turner MG (ed) Landscape heterogeneity and disturbance. Ecological studies, vol 64. Springer-Verlag, New York, pp 59–83
Kovacic DA, Dyer MI, Cringan AT (1985) Understory biomass in ponderosa pine following mountain pine beetle infestations. For Ecol Manag 13:53–67
Kueppers LM, Southon J, Baer P, Harte J (2004) Dead wood biomass and turnover time, measured by radiocarbon, along a subalpine elevation gradient. Oecologia 141:641–651
Laiho R, Prescott CE (1999) The contribution of coarse woody debris to carbon, nitrogen, and phosphorus cycles in three Rocky Mountain coniferous forests. Can J For Res 29:1592–1603
Lundquist JE (2007) The relative influence of diseases and other small-scale disturbances on fuel loading in the black hills. Plant Dis 91(2):147–152. doi:10.1094/pdis-91-2-0147
Maguire DA (1994) Branch mortality and potential litterfall from Douglas-fir trees in stands of varying density. For Ecol Manag 70:41–53
Marra JL, Edmonds RL (1996) Coarse woody debris and soil respiration in a clearcut on the Olympic Peninsula, Washington, USA. Can J For Res 26:1337–1345
Martin RE, Anderson HE, Boyer WD, Dieterich JH, Hirsch SN, Johnson VJ, McNab WH (1979) Effects of fire on fuels. USDA forest service, General Technical Report WO-13, Washington, DC, 64 pp
Matson E, Bart D (2013) Interactions among fire legacies, grazing and topography predict shrub encroachment in post-agricultural páramo. Landsc Ecol 28(9):1829–1840. doi:10.1007/s10980-013-9926-5
McKenzie D, Kennedy MC (2011) Scaling laws and complexity in fire regimes. In: McKenzie D, Miller C, Falk DA (eds) The landscape ecology of fire. Springer, Dordrecht, pp 27–51
McKenzie D, Miller C, Falk DA (eds) (2011) The landscape ecology of fire. Springer, Dordrecht
McKenzie D, Shankar U, Keane RE, Stavros EN, Heilman WE, Fox DG, Riebau AC (2014) Smoke consequences of new wildfire regimes driven by climate change. Earth's Fut:n/a-n/a. doi:10.1002/2013ef000180
Means JE, Cromack K, MacMillan C (1985) Comparison of decomposition models using wood density of Douglas-fir logs. Can J For Res 15:1092–1098
Meentemeyer V (1978) Macroclimate and lignin control of litter decomposition rates. Ecology 59:465–472
Metz MR, Frangioso KM, Meentemeyer RK, Rizzo DM (2010) Interacting disturbances: wildfire severity affected by stage of forest disease invasion. Ecol Appl 21(2):313–320. doi:10.1890/10-0419.1
Metz MR, Varner JM, Frangioso KM, Meentemeyer RK, Rizzo DM (2013) Unexpected redwood mortality from synergies between wildfire and an emerging infectious disease. Ecology 94(10):2152–2159. doi:10.1890/13-0915.1
Miller RF, Rose JA (1999) Fire history and western juniper encroachment in sagebrush steppe. J Range Manag 1999:550–559
Miller C, Urban DL (2000) Connectivity of forest fuels and surface fire regimes. Landsc Ecol 15:145–154
Millar CS (1974) Decomposion of coniferous leaf litter. In: Dickinson CH, Pugh GJF (eds) Biology of plant litter decomposition, vol 1. Academic, London, pp 105–129
Mitchell SJ (2013) Wind as a natural disturbance agent in forests: a synthesis. Forestry 86(2):147–157. doi:10.1093/forestry/cps058
Moreira F, Rego F, Ferreira P (2001) Temporal (1958–1995) pattern of change in a cultural landscape of northwestern Portugal: implications for fire occurrence. Landsc Ecol 16(6):557–567. doi:10.1023/a:1013130528470

Muller RN (2003) Landscape patterns of change in coarse woody debris accumulation in an old-growth deciduous forest on the Cumberland Plateau, southeastern Kentucky. Can J For Res 33:763–769
Noble IR, Slatyer RO (1980) The use of vital attributes to predict successional changes in plant communities subject to recurrent disturbances. Vegetatio 43:5–21
Olson J (1963) Energy storage and the balance of the producers and decomposers in ecological systems. Ecology 44:322–331
Page W, Jenkins MJ (2007) Mountain pine beetle-induced changes to selected lodgepole pine fuel complexes within the Intermountain Region. For Sci 53(4):507–520
Parker TJ, Clancy KM, Mathiasen RL (2006) Interactions among fire, insects, and pathogens in coniferous forests of the interior western United States and Canada. Agric For Entomol 8:167–189
Parsons RA, Mell WE, McCauley P (2010) Linking 3D spatial models of fuels and fire: effects of spatial heterogeneity on fire behavior. Ecol Model 222(3):679–691
Pearson JA (1987) Biomass and nutrient accumulation during stand development in Wyoming lodgepole pine forests. Ecology 68:1966–1977
Peters DPC, Mariotto I, Havstad KM, Murray LW (2006) Spatial variation in remnant grasses after a grassland-to-shrubland state change: implications for restoration. Rangeland Ecol Manag 59:343–350
Pickett STA, White PS (1985) The ecology of natural disturbance and patch dynamics. Academic San Diego
Prescott CE, Zabek LM, Staley CL, Kabzems R (2000) Decomposition of broadleaf and needle litter in forests of British Columbia: influences of litter type, forest type, and litter mixtures. Can J For Res 30:1742–1750
Prescott CE, Hope GD, Blevins LL (2003) Effect of gap size on litter decomposition and soil nitrate concentrations in a high-elevation spruce-fir forest. Can J For Res 33:2210–2220
Reich RM, Lundquist JE, Bravo VA (2004) Spatial models for estimating fuel loads in the Black Hills, South Dakota, USA. Int J Wildland Fire 13:119–129
Reinhardt E, Scott J, Gray K, Keane R (2006) Estimating canopy fuel characteristics in five conifer stands in the western United States using tree and stand measurements. Can J For Res 36(11):2803–2814. doi:10.1139/X06-157
Reinhardt ED, Keane RE, Calkin DE, Cohen JD (2008) Objectives and considerations for wildland fuel treatment in forested ecosystems of the interior western United States. For Ecol Manag 256(12):1997–2006. doi:10.1016/j.foreco.2008.09.016
Schoennagel TL, Veblen TT, Romme WH (2004) The interaction of fire, fuels, and climate across rocky mountain landscapes. Bioscience 54(7):651–672
Schoennagel T, Veblen TT, Negron JF, Smith JM (2012) Effects of mountain pine beetle on fuels and expected fire behavior in lodgepole pine forests, Colorado, USA. PLoS ONE 7(1):e30002
Six DL, Adams J (2007) White pine blister rust severity and selection of individual whitebark pine by the mountain pine beetle (Coleoptera: Curculionidae, Scolytinae). J Entomol Sci 42(3):345–353
Sollins P (1982) Input and decay of coarse woody debris in coniferous stands in western Oregon and Washington. Can J For Res 12:18–28
Sollins P, Cline SP, Verhoeven T, Sachs D, Spycher G (1987) Patterns of log decay in old growth Douglasfir forests. Can J For Res 17:1585–1595
Spies TA, Franklin JF, Thomas TB (1988) Coarse woody debris in Douglas-fir forests of western Oregon and Washington. Ecology 69:1689–1702
Stohlgren TJ (1988) Litter dynamics in two Sierran mixed conifer forests. II. Nutrient release from decomposing litter. Can J For Res 18:1136–1144
Stump LM, Binkley D (1993) Relationships between litter quality and nitrogen availability in Rocky Mountain forests. Can J For Res 23:492–502
Sturtevant B, Zollner P, Gustafson E, Cleland D (2004) Human influence on the abundance and connectivity of high-risk fuels in mixed forests of northern Wisconsin, USA. Landsc Ecol 19(3):235–253. doi:10.1023/B:LAND.0000030420.87498.f7

Swift MJ, Heal OW, Anderson JM (1979) Decomposition in terrestrial ecosystems, vol 5. University of California Press, Berkeley. ISBN:0520040015

Taylor BR, Prescott CE, Parsons WFJ, Parkinson D (1991) Substrate control of litter decomposition in four Rocky Mountain conifer forests. Can J Bot 69:2242–2250

Thaxton JM, Platt WJ (2006) Small-scale fuel variation alters fire intensity and shrub abundance in a pine savanna. Ecology 87(5):1331–1337. doi: 10.1890/0012-9658(2006)87[1331:SFVAFI]2.0.CO;2

Theobald D (2013) A general model to quantify ecological integrity for landscape assessments and US application. Landsc Ecol 28(10):1859–1874. doi:10.1007/s10980-013-9941-6

Townsend DE, Fuhlendorf SD (2010) Evaluating relationships between spatial heterogeneity and the biotic and abiotic environments. Am Midl Nat 163(2):351–365. doi:10.1674/0003-0031-163.2.351

Trofymow JA (1991) Annual rates and elemental concentrations of litter fall in thinned and fertilized Douglas-fir. Can J For Res 21:1601–1615

Turner M, Bratton SP (1987) Fire, grazing, and the landscape heterogeneity of a Georgia barrier island. In: Turner MG (ed) Landscape heterogeneity and disturbance. Ecological studies, vol 64. Springer, New York, pp 85–101. doi:10.1007/978-1-4612-4742-5_5

Turner MG, Dale VH, Everham III EH (1997) Fires, hurricanes, and volcanoes: comparing large disturbances (Large disturbances are heterogeneous spatially and not as catastrophic as they may initially appear). BioScience 47(11):758–768

Ulanova NG (2000) The effects of windthrow on forests at different spatial scales: a review. For Ecol Manag 135(1–3):155–167. doi:http://dx.doi.org/10.1016/S0378-1127(00)00307-8

Vihnanek RE, Balog CS, Wright CS, Ottmar RD, Kelly JW (2009) Stereo photo series for quantifying natural fuels. Post-hurricane fuels in forests of the Southeast United States, vol 12. U.S. Department of Agriculture, Forest Service, Pacific Northwest Research Station, General technical report PNW-GTR-803 Portland, OR USA, 53 pp

Vogt KA, Grier CC, Vogt DJ (1986) Production, turnover, and nutrient dynamics of above-and belowground detritus of world forests. Adv Ecol Res 15:303–377

Waring RH, Schlesinger WH (1985) Forest ecosystems concepts and management. Academic, San Diego

Waring RH, Running SW (1998) Forest ecosystems: analysis at multiple scales, 2nd edn. Academic, San Diego

White DE, Atzet T, Martinez PA (2004) Relationship of historic fire regimes to dead wood components in white fir forests of southwestern Oregon. In: Fire in temperate, boreal and montane ecosystems: Proceedings of the 22nd Tall Timbers Fire Ecology Conference. Tall Timbers Research Station, Tallahassee, FL USA, pp 117–124

Wright KH, Lauterback PG (1958) A 10-year study of mortality in a Douglas-fir sawtimber stand in Coos and Douglas Counties, Oregon. Research Paper PNW-27 USDA Forest Service Pacific Northwest Forest and Range Experiment Station, Portland, OR USA

Yavitt JB, Fahey TJ (1982) Loss of mass and nutrient changes of decaying woody roots in lodgepole pine forests, southeastern Wyoming. Can J For Res 12:745–752

Zimmermann TG, Neuenschwander LF (1984) Livestock grazing influences on community structure, fire intensity, ad fire frequency within the Douglas-fir/Ninebark habitat type. J Range Manag 37(2):104–110

Section II
Fuel Applications

Chapter 7
Fuel Classifications

Science is the systematic classification of experience
Philosopher George Henry Lewes

7.1 Introduction

Wildland fire scientists and managers use classifications of fuelbeds for a number of reasons. Most importantly, classifications provide a means to easily enter fuelbed properties into fire management software. Fire managers often have insufficient resources to directly measure or sample fuel component characteristics in the field, so using a classification to quantify fuel characteristics is an appealing option. Second, many use classifications to communicate fuelbed characteristics to other professionals because most fuelbeds are highly complex and diverse (Chap. 2), and this complexity often limits effective technical exchange, especially in operational fire management planning and tactical firefighting. Third, the categories in some fuel classifications may be used as mapping units in the development of digital fuel maps over large areas (Chap. 9). Finally, some classifications can be used in the field as an alternative fuel inventory and monitoring protocol for assessing fuel loadings (Sikkink et al. 2009) (Chap. 8).

Classification is often defined as the process in which objects are recognized, differentiated, and understood. In this chapter, fuel classification is defined as the process of identifying unique fuelbeds and quantifying their component attributes. People differentiate fuelbeds in a number of ways. Some assume vegetation serves as an acceptable surrogate for differentiating fuelbeds, so they use vegetation classifications as de facto fuel classifications (Keane et al. 2013). Others classify fuelbeds by the way they might burn in a severe fire (Burgan 1987; Hornby 1935). While some subjectively evaluate the representativeness of a fuelbed through field reconnaissance (Ottmar et al. 2007), others use extensive field data to systematically classify fuelbeds using advanced statistical techniques (Lutes et al. 2009). Fuel classifications may use any number of variables to describe and quantify fuel component attributes, such as heat content, mineral content, and particle density, depending on the fire software application, but the most common variable used across fire management classifications is fuel loading (Weise and Wright 2014).

R. E. Keane, *Wildland Fuel Fundamentals and Applications*,
DOI 10.1007/978-3-319-09015-3_7

7.2 Classification Approaches

Several fuel classifications are currently used by land management agencies across the globe, and most of these systems appear quite similar because they have comparable categories, components, and description variables (Anderson 1982; Keane 2013; Sandberg et al. 2001; Weise and Wright 2014). The main distinction between most existing fuel classification systems is in the approaches used to create them (Keane 2013). Although it would be much easier if there was only one fuel classification for all fire science and management applications, multiple fuel classification systems exist today because each fire modeling system requires a specific set of fuel inputs and its own unique classification input scheme. Fire behavior fuel classifications, for example, include fuel component attributes, such as fuel depth, that may not be needed in fire effects prediction systems.

Effective biological classifications are designed to be systematic (well organized), practical (easily identified using a key), singular (uniquely identifies a class), and comprehensive (the key can be used across a broad range of fuelbeds). This usually implies that the classes that comprise them are mutually exclusive, and a change in the value of an attribute of one class usually affects the values of the same attribute in other classes (Gauch and Whittaker 1981). However, many of today's fuel classifications were not created using systematic classification procedures that group fuelbeds based on statistical and ecological differences. Because of this, the fuel classifications in this chapter will be summarized by the four broad approaches used to create them: (1) association, (2) opportunistic, (3) classification, and (4) abstraction (Table 7.1). Of course, some of the fuel classifications presented as examples were created using a combination of approaches.

7.2.1 Association

Many have associated or linked fuel component information, such as loading, to the categories of other extant classifications commonly used in natural resource management (Keane 2013). This is often accomplished by summarizing field-collected fuels data by extant classification categories. For example, Reinhardt et al. (1997) average field-measured fuel loadings for eight fuel components across the vegetation-based categories of both the Eyre (1980) forest cover type classification and the Shiflet (1994) range cover type classification to facilitate input to the First-Order Fire Effects Model (FOFEM). In Canada, Hawkes et al. (1995) assigned fuel loadings to various categories of vegetation and timber type classifications, and the Canadian Fire Behavior Prediction System contains fuel input types that are associated with major forest vegetation types (FCFDG 1992). Poulos et al. (2007) created vegetation composition and structure layers from environmental gradients, satellite imagery, and forest inventory data, then scaled fuels information to the resultant biophysical classification for Texas fuelbeds. The fuel type group classification was created by summarized Forest Inventory and Analysis georeferenced fuels data by forest type groups (Keane et al. 2013).

Table 7.1 Comparison of the four approaches used to develop wildland fuel classifications for fire management

Approach	Description	Advantages	Disadvantages	Examples
Association	Fuels information is assigned to categories in extant classifications	Existing vegetation classifications and maps can be used to describe fuels; most existing classifications are widely used, well accepted, and used in all phase of land management; contain comprehensive keys to uniquely identify classes in the field	Fuel properties are often uncorrelated to vegetation categories; extant classifications are often too broad to describe subtle changes in fuels; fuel properties can be redundant across the classes; difficult to alter extant classifications to account for differences in fuel attributes	US: FOFEM SAF/SRM Cover types (Reinhardt et al. 1997), FCCS (Ottmar et al. 2007) Canada: FBP (Hawkes et al. 1995) China: Wu et al. (2011) Russia: Volokitina and Sofronov (2000)
Opportunistic	Unique fuelbeds are identified and sampled in the field and the fuelbed is added as another category in the classification	Represent real fuelbeds; designed for fine scales; easily add new fuel categories; easily add new fuel components	Data intensive; classes can be highly redundant; infinite number of classes makes it difficult to learn and use; many fuel types are missing because they are not yet sampled	US: FCCS (Ottmar et al. 2007), photo series (Vihnanek et al. 2009)
Classification	Fuel data are clustered into similar groups that are mutually exclusive using statistical techniques	Can control variation and limit redundancy across classes; comprehensive keys can be developed to identify classes in field; can design classification for any scale, area, or fuel type; easy to learn and use	Data intensive and many fuelbeds may not be represented in data; difficult to add new fuel types and new fuel components into the classification; classification is complex and difficult to understand	US: FLM (Lutes et al. 2009), Crown assessment (Fahnestock 1970) Greece: Dimitrakopoulos (2001) Australia: Gould et al. (2011)
Abstraction	Fuels inputs to fire models are adjusted to match observed fire behavior, and the adjusted fuel information becomes a category in the classification	Match the resolution of the fire models; widely used and accepted in fire management; training widely available for managers	Do not represent real fuel fuelbeds; can't directly use to estimate fuel loadings; difficult to create, use, and interpret; no keys are available to identify classes in the field; only useful for use in the fire behavior model for which they were developed	US: FBFMs: Anderson (1982), Scott and Burgan (2005) Greece: Dimitrakopoulos (2002)

There are many advantages to linking fuels to vegetation-based classifications that make this approach quite attractive to a number of researchers and managers (Bailey and Mickler 2007). Many vegetation and natural resource classifications are well known to fire managers and have a long history of use in land management because they are easy to learn and contain proven keys for quick and objective identification of vegetation categories in the field. Vegetation characteristics used in classification keys, such as composition, structure, and successional stage, are easily identified in the field with minimal training. Moreover, a vast array of ancillary land management analyses can be done by linking vegetation information with fuels data, such as predicting future fuel conditions using vegetation succession models (Davis et al. 2009), linking canopy fuels with surface fuels (Keane et al. 2006), creating fuel maps (Reeves et al. 2006), and prioritizing areas for fuel treatment (Hessburg et al. 2007). Finally, additional fuel components and characteristics can be added with little effort; canopy fuels, for example, can be summarized by vegetation type along with surface fuel loadings.

There are some major problems with the association method of linking fuel characteristics to existing classification categories that might limit the application of this approach in the future (Table 7.1). First and foremost, fuel characteristics are rarely correlated to vegetation attributes and categories, especially at fine scales, because they also depend on decomposition and disturbance (Chap. 6) (Keane et al. 2012b; Keane and Gray 2013). Brown and Bevins (1986) found that fuel loadings did not correlate with cover type or habitat type and speculated that stand disturbance history had more influence on fuelbed loadings than vegetation. One reason for this lack of relationship between fuels and vegetation might be that vegetation attributes, such as species cover and height, vary at coarser scales than wildland fuels (Chap. 6). Wildland fuel loadings are also highly variable across a vegetation type category (Chap. 6). As a result, many disparate fuelbeds may be represented within one vegetation type, and conversely, many vegetation types may have the same fuelbed description. This redundancy is also related to the fact that the resolutions of most vegetation classifications (e.g., species taxa) do not match the resolution of those fuelbed characteristics that foster unique fire behavior and effects (fine-scale fuel components) (Keane et al. 2012a). For these reasons, vegetation-based fuel classifications often have poor accuracies and low precisions (Keane et al. 2013). Accuracies of the vegetation classifications for which fuels are associated do not reflect the true accuracy of the fuel information. For example, a 90 % accuracy of a vegetation map does not translate into 90 % accuracy for the fuels data. The associated fuels information must be compared with field-collected fuel data to determine fuelbed accuracy, and often, these analyses show poor agreement (Keane et al. 2013). Moreover, since fuel component properties are independently averaged across somewhat broad vegetation categories, the resultant set of fuel component properties may represent a summarized fuelbed that may be rare.

Another problem with the association approach is that it is difficult to refine the fuel descriptions to improve classification accuracies. If classified fuel loading accuracies are low, as is often the case, there is little recourse to improve the accuracy without changing the original vegetation classifications by adding, modifying, or deleting categories, or by adding additional classifications to the already complex associative approach (e.g., combine a classification of stand structure with a cover type classification). The addition of new classifications or classes exponentially increases the amount of fuel data needed to cover all combinations of the merged classifications; so many combinations might be missing valuable fuel data to quantify fuel information.

7.2.2 Opportunistic

In the opportunistic approach to fuel classification, unique fuelbeds are subjectively identified in the field and selected as a new category to include in the classification based on their representativeness for a region, vegetation type, or fuel type. The newly identified fuelbed becomes a new class in the classification once the fuel component properties are measured and assigned to this fuelbed. Keane (2013) called this a "bottom-up" indirect classification approach where there are an infinite number of classes possible in this ever-expanding classification method.

Two fuel classifications provide excellent examples of this opportunistic approach: the photo series (Chap. 8) and the Fuel Characteristics Classification System (FCCS). In both, new and unique fuelbeds can be added as they are identified by managers, scientists, and resources specialists in the field for local, regional, or national applications (Berg 2007). When new fuelbeds are sampled, the resultant data become attributes of the new class in the classification (Riccardi et al. 2007b). The photo series is a set of photographs of fuelbeds where fuel component loadings have been measured (Fig. 7.1). These photographs are usually described and stratified by vegetation characteristics, such as cover type or species composition. Each photo in the series becomes a category in the classification and many have used photo series photos to describe and quantify fuel characteristics (Keyes 2002). The FCCS is a more formal adoption of an opportunistically derived fuel classification (Ottmar et al. 2007). In the FCCS, unique fuelbeds are identified, either in the field or office, and then directly or indirectly sampled to populate a database that links

Fig. 7.1 A picture from the Fischer (1980) photo series

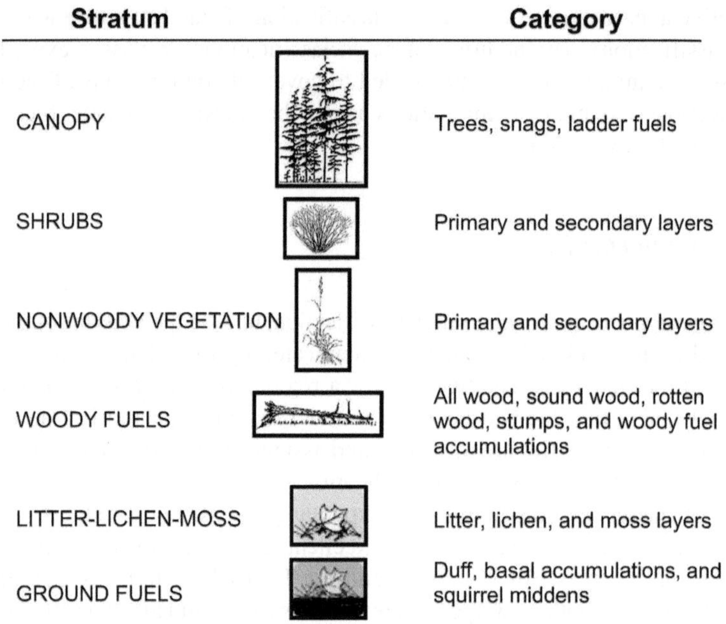

Fig. 7.2 A general description of the elements in the FCCS. (Ottmar et al. 2007)

fuel component properties with the identified FCCS "fuelbed." This fuelbed then becomes a category in the classification. The system uses ecoregion, stand structure, and site history classification variables to identify fuelbeds in the field (Riccardi et al. 2007a). FCCS is also somewhat special in that it also contains its own fire behavior model tuned for the FCCS fuel components (Sandberg et al. 2007).

The advantage of developing opportunistic classifications is that new fuel components and properties can be added to the classification with little effort. The FCCS has quantified over 20 fuel properties for several fuel components in each fuelbed in the classification (Fig. 7.2). Opportunistic classifications can be used to represent fuels at any scale; FCCS classifications have been developed for small areas, such as plots and treatment units, and for large regions, such as the entire USA (McKenzie et al. 2007). These classifications are also easy to understand and build, and they can be modified and revised by anyone with any level of experience. In addition, the classes represent actual fuelbeds that are extensively documented in the field.

There are some shortcomings in the opportunistic approach that may limit their application. Few opportunistic classifications are able to consistently and uniquely identify a fuelbed in the field (Ottmar et al. 2007). Most rely on the expertise of the fuel sampler to match the observed fuelbed conditions to the categories in the classification, or to identify a class based on the ancillary vegetation and site classification criteria used to describe the fuelbed (e.g., photo series). The FCCS, for example, does not contain a key to directly identify a fuelbed from fuelbed characteristics.

Instead, it uses a set of ecological descriptions mostly based on vegetation and stand history to aid in fuelbed identification (Ottmar et al. 2007). As a result, there is often redundancy across many fuel classification categories; the properties of one fuelbed may be quite similar to other fuelbeds sampled in another part of the country or for another vegetation type, especially for the fine woody debris components. Linking opportunistic classification categories to spatial data layer attributes is also problematic because it is difficult to consistently validate an assigned class in the field because there is no fuel classification key. Another problem is that since the variation across fuelbeds is not incorporated into the classification design, there can be an infinite number of possible categories (fuelbeds), and conversely, there can be many locally relevant fuelbeds that are missing in the final classification. Keane et al. (2006), for example, mapped FCCS categories across central Utah but found that over 30 % of the land area had vegetation attributes that did not match sampled FCCS classes. This issue makes opportunistic classifications somewhat difficult to learn because it is always changing and new classes are always being added.

7.2.3 Classification

Classification, as previously mentioned, is the process of systematically and comprehensively clustering items (fuelbeds) into unique groups based on selected attributes—mainly loading by fuel components. Usually, this involves numerical clustering and complex statistical techniques that attempt to directly identify unique groups based on the variation of the attributes selected to develop the classification (Gauch and Whittaker 1981; Orloci 1967). Once unique groups are identified, a comprehensive key based on the analysis variables (e.g., loading) can be devised to objectively identify the classification category for a field-assessed observation. This approach partitions the variation in the field data to reduce redundancy and produce a singular classification.

Few existing fuel classifications were built using this direct, top-down classification approach. In perhaps the first effort at directly classifying fuels, Fahnestock (1970) developed two keys that evaluated various fuel attributes, including particle size, compactness, vertical position, and horizontal continuity, to key to unique spread rate and crowning potential classes. Dimitrakopoulos (2001) created a fuels classification for Greece by clustering flammability variables, such as heat content, ash content, and particle density, into unique groups using hierarchical cluster analysis and canonical discriminant analysis for Mediterranean shrublands. The fuel loading models (FLMs) of Lutes et al. (2009) is distinctive in that field-collected fuel loading data were used to simulate smoke emissions and soil heating, and these simulation results, along with loading, were used to create unique classes using advanced clustering and then a unique key was created using regression tree analyses. As a result, this classification effectively integrated the resolution of the fire models for which the FLMs would eventually be used into the classification design (Fig. 7.3).

An advantage of the direct classification approach is that resultant classifications are fully supported by the data that were used to create them, and therefore,

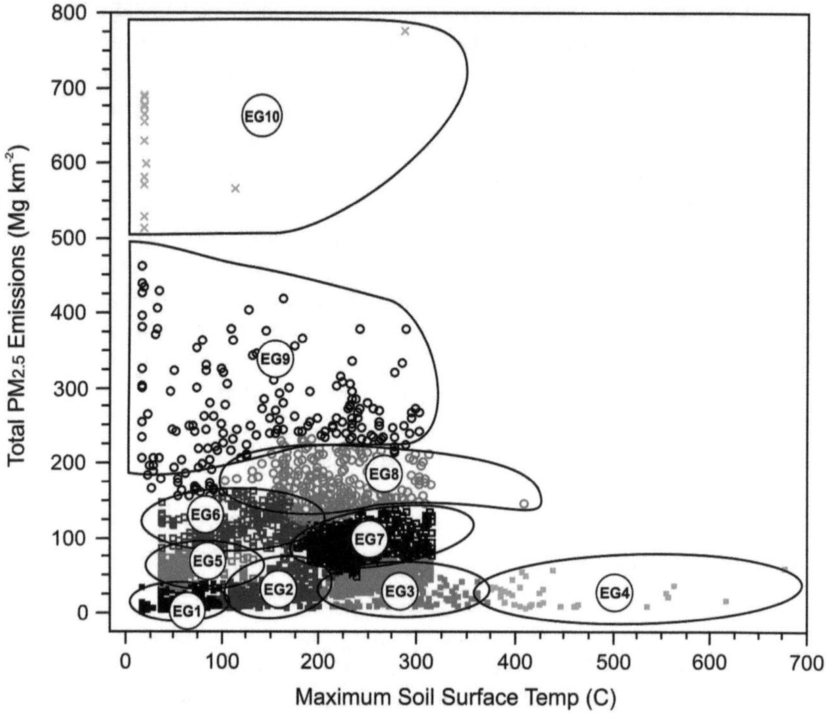

Fig. 7.3 The classification diagram showing the clustering of fire effects groups (e.g., EG1) on gradients of smoke emissions production and soil heating. These fire effects groups were then divided into finer groups to create the FLMs. (Lutes et al. 2006)

represent actual fuelbeds with measured loadings. As such, these classifications can be used as (1) inventory techniques to quantify fuel characteristics (Sikkink et al. 2009); (2) descriptors of unique fuel types to facilitate communication between managers, scientists, and other professionals (Sandberg et al. 2001); and (3) map units in fuel mapping efforts (Keane et al. 2001). Effective classified fuel systems contain dichotomous keys that can uniquely identify a class on the ground based on qualities of the fuelbed (Sikkink et al. 2009). The loading information for a classified category can be used in fire applications, such as simulating fire effects and validating fuel maps, and the variability of loadings within a category can be incorporated into the analyses. And since statistical classifications have low redundancy between classes, class attributes may be used for quantifying loading in fire models, as a field inventory technique (Chap. 7), and for identifying possible thresholds in fire behavior and effects modeling (Lutes et al. 2009).

Directly classified fuel classifications, such as FLMs, also have drawbacks. All fuel classifications, and especially those developed from direct classification techniques, require extensive data sets to fully represent the diversity of fuelbeds in the analysis. As a result, the depth, scope, and quality of the data sets used to create the classification system are rarely comprehensive enough to represent all possible fuelbeds that exist across the target area. While FLMs were developed using extensive data collected across the entire USA, the analysis data set was missing critical data from several major US fuelbeds that were unsampled at the time of FLM development, including many non-forest rangeland types, and therefore, these categories are missing in the classification (Lutes et al. 2009). Another limitation is that the parameters used in the clustering algorithms, such as the desired number of clusters, have a major influence on the classification, yet they are often subjectively estimated based on the objectives of the analysis. Finally, it is quite difficult to modify, add, or remove new categories or components as new data become available without completely redoing the entire classification.

7.2.4 Abstraction

Some fuel classifications were created using abstraction where the qualities of a fuelbed are related to abstract evaluations of fire behavior (Muraro 1965). Hornby (1936), for example, subjectively described western US fuelbeds using two fire behavior attributes: resistance to fire control and fire spread (Chap. 1). Most US fire behavior predictions systems were built using the Rothermel (1972) model, and the fuel classifications used as inputs to this model are often called fire behavior fuel models (FBFMs) that are essentially abstractions of expected fire behavior. Each FBFM is described by a set of fuel characteristics (e.g., loading, SAVR, mineral content, heat content) for each of the input fuel components required by the fire behavior modeling systems (Burgan and Rothermal 1984). However, the FBFM fuel characteristics are quantified to represent "expected" fire behavior and, as such, can't be used to describe actual fuel characteristics. To create FBFMs, fuel input parameters for each FBFM, including loading, are adjusted to reflect realistic fire behavior under known fuel moisture and weather conditions by comparing model results with observed fire behavior or expert opinion (Burgan 1987). This is because the inherent complexity of the quasi-mechanistic Rothermel (1972) fire behavior algorithm makes it difficult to predict realistic fire behavior from actual fuel loadings (Burgan 1987). As a result, a somewhat complicated procedure has been developed to create new FBFM models, called "custom" fuel models, where fuel loadings and other fuelbed characteristics need to be adjusted to achieve a realistic and believable fire simulations based on observations of fire behavior in the field. As a result, FBFMs are actually classifications of expected fire behavior. They were included in this chapter because they are perhaps the most used fuel classification in fire management. FBFMs have been used in the USA for over 30 years, and they have been broadly accepted by managers as a viable method of describing fuels for fire behavior modeling. The development and use of FBFMs are taught to fire managers in a wide variety of fire management courses throughout the world.

Most abstract fuel description systems today are FBFMs created for use in fire behavior applications that contain the Rothermel (1972) spread model as implemented in BEHAVE (Andrews 2008) and FARSITE (Finney 1998) systems. In the USA, the most commonly used FBFM classifications are the (1) 13 FBFMs described by Anderson (1982), (2) 40+ models of Scott and Burgan (2005), and (3) 20 fire danger fuel models used in the National Fire Danger Rating System (Deeming et al. 1977). Others have created new sets of custom FBFMs to these classifications. Reich et al. (2004), for example, created several new BEHAVE custom fuel models using field loading data that were then mapped to a South Dakota US landscape, and Cheyette et al. (2008) created custom fuel models for the wildland urban interface lands around Anchorage, Alaska, using a supervised vegetation-based classification of 13 cover types. In Greece, Dimitrakopoulos (2002) created seven FBFMs by synthesizing fuel data from 181 natural fuel complexes described by vegetation. In Corsica, Santoni et al. (2011) developed two fuel models for a spatially explicit fire model built to simulate fire behavior for maquis and juniper shrublands. To evaluate fire hazard in Portugal, Fernandes (2009) developed a suite of 19 fuel models based on the dominant vegetation structures and complexes in mainland Portuguese forests.

The main advantage in creating abstract fuel description systems is that, ideally, the resolution of fuel classes (FBFMs) match the resolution of the fire models for which the classes will be used as inputs. Another words, each FBFM represents a major change in predicted fire behavior in the Rothermel (1972) model. This means that the uncertainty and error in model predictions may be minimized from inaccurate and inappropriate fuel inputs because the fuel models were calibrated to actual fire behavior observations (Burgan 1987). Another advantage is that new custom fuel models can be developed for unique local situations or for broad use across large regions (Burgan and Hardy 1994).

The biggest drawback to the abstraction classification approach and their products, such as FBFMs, is that without prior knowledge of fire behavior in local fuel conditions, it is nearly impossible to accurately and consistently identify, use, and interpret most of the abstract classes. Identification of FBFMs in the field, for example, is highly subjective because it is based on an individual's perception of how fire will burn the fuelbed under severe weather conditions, rather than on actual measurements of fuel loadings. There are no standardized keys to consistently identify FBFMs for either the Anderson (1982) or Scott and Burgan (2005) FBFM classification systems. Because abstract classifications are inherently subjective and difficult to use, most fuel mapping efforts based on abstract classification products must rely on expert knowledge and past experience (Keane and Reeves 2011). FBFMs are also difficult to create because their development requires a delicate balance of parameter adjustments to match observed fire behavior with fire weather and fuel properties that should only be done by experienced analysts and fire managers (Burgan 1987). These limitations may preclude the use of FBFMs in the future as new fire behavior simulation models are developed, as novel fuelbeds are created from innovative fuel treatments, and as abundant fuel input data become available for describing fuelbeds.

Abstract fuel classifications can only be used for fire behavior prediction and are rarely used in other areas of fire and land management. FBFMs, for example, don't include loadings for some major fuel components, such as logs and duff, which are critical for computing smoke emissions, simulating post-frontal combustion, and evaluating wildlife habitat. FBFMs can only be used in the fire behavior model for which they were created; it is inappropriate to use existing fire behavior or danger fuel models in other fire simulation systems (Alexander 2013; Sandberg et al. 2007). Similar to opportunistic classification approaches, there can be an infinite number of abstractions to account for an infinite number of possible fire behaviors, making FBFMs that represent unique fire behaviors difficult to build, especially given the coarse resolution of the fire models. And because FBFMs indirectly represent the resolution of the fire behavior prediction systems, it is difficult to evaluate the effect that subtle changes in fuel characteristics brought about by fuel treatments have on fire behavior, especially if there are small changes in fuel loadings that are too fine for the resolution of the FBFM.

7.3 Challenges

Classifying wildland fuelbeds has always been difficult because of the highly variable composition, distribution, and arrangement of fuel particles in space and the dynamic changes in particle characteristics over time (Chaps. 2, 3, and 5). Spatial and temporal variability of fuel properties directly influences fire behavior (Parsons et al. 2010; Bachmann and Allgower 2002), controls fire effects (Reinhardt et al. 2001), confounds fuel sampling (Keane and Gray 2013), confuses mapping efforts (Keane et al. 2001), and complicates fuel classification (Keane 2013). Fuel properties are highly variable across space and can even be highly variable within individual fuel particles (Keane et al. 2012b). This variability is scale dependent with variability of smaller fuel particles distributed over smaller scales than large fuels (e.g., twigs vary at smaller scales than logs). Any fuel classification system that does not incorporate this variability into its design may be highly redundant and ineffective for some fire applications.

Fire managers and researchers are often frustrated by all these seemingly redundant classifications and may desire a single fuel description system that can be used across all software platforms and prediction systems. This would simplify fuel sampling, mapping, and input into the numerous fire management applications. This chapter presents several reasons why today's fuel classifications often have insufficient scope, quality, resolution, and accuracy to serve as the primary fuel classification in fire management. Several major advances in technology and research need to be made before a universal fuel description system can be created. It will be difficult to develop any new fuel description system without knowing what the new fire models need for fuels inputs. While the next generations of fire behavior and effects simulation models are being developed, it is critical that both new fuel classification systems be built to balance ecological understanding of fuel dynamics with both old and new input model requirements. It is also critical that future fire behavior models be implemented in three dimensions (3D) to account for the spatial distributions of fuel and its effect on fire behavior, especially those models used in fire research (Krivtsov et al. 2009). And, each of these characteristics must have an associated sampling method for accurate quantification, and these methods must account for the wide diversity of fuel particles comprising the fuelbed (Chap. 8). And last, the development of new comprehensive fuel classifications will need high quality data across large geographical areas, diverse ecosystems, and complex fuelbeds to ensure effective and robust applications (Conard et al. 2001).

References

Alexander ME (2013) Part 4: The science and art of wildland fire behaviour prediction. In: Scott AC, Bowman DMJS, Bond WJ, Pyne SJ, Alexander ME (eds) Fire on Earth: An Introduction. Wiley-Blackwell, Chichester

Anderson HE (1982) Aids to determining fuel models for estimating fire behavior. USDA Forest Service Intermountain Research Station, General Technical Report INT-122, Ogden, pp 22

Andrews PL (2008) BehavePlus fire modeling system, version 4.0: Variables. USDA Forest Service Rocky Mountain Research Station, General Technical Report RMRS-GTR-213WWW, Fort Collins, pp 107

Bachmann A, Allgower B (2002) Uncertainty propagation in wildland fire behaviour modelling. International J Geogr Inf Sci 16(2):115–127. doi:10.1080/13658810110099080

Bailey AD, Mickler R (2007) Fine scale vegetation classification and fuel load mapping for prescribed burning. In: Butler BW, Cook W (eds) The fire environment-innovations, management, and policy, Destin, FL. USDA Forest Service Rocky Mountain Research Station, pp 261–270

Berg E (2007) Characterizing and classifying complex fuels—A new approach. Can J For Res 37(12):2381–2382

Brown JK, Bevins CD (1986) Surface fuel loadings and predicted fire behavior for vegetation types in the northern Rocky Mountains. U.S. Department of Agriculture, Forest Service, Intermountain Forest and Range Experiment Station, Research Paper INT-358, Ogden, pp 9

Burgan RE (1987) Concepts and interpreted examples in advanced fuel modeling. USDA Forest Service, Intermountain Research Station General Technical Report INT-238, Ogden, pp 40

Burgan RE, Rothermal RC (1984) BEHAVE: fire behavior prediction and fuel modeling system–FUEL subsystem. USDA Forest Service, Intermountain Research Station General Technical Report INT-167, Ogden, pp 126

Burgan RE, Hardy CC (1994) Ground truthing a national AVHRR based vegetation fuels map. In: Proceedings of the 12th Conference on Fire and Forest Meteorology, Society of American Foresters, Bethesda, Maryland, pp 428–436

Cheyette D, Rupp TS, Rodman S (2008) Developing Fire Behavior Fuel Models for the WildlandUrban Interface in Anchorage, Alaska. Western J Appl For 23:149–155

Conard SG, Hartzell T, Hilbruner MW, Zimmerman GT (2001) Changing fuel management strategies – The challenge of meeting new information and analysis needs. Intl J Wildland Fire 10(4):267–275. doi:http://dx.doi.org/10.1071/WF01027

Davis B, Van Wagtendonk JW, Beck J, van wagtendonk K (2009) Modeling fuel succession. Fire Manag Today 69(2):18–21

Deeming JE, Burgan RE, Cohen JD (1977) The National Fire Danger Rating System—1978. USDA Forest Service Intermountain Forest and Range Experiment Station, General Technical Report INT-39, Ogden, pp 63

Dimitrakopoulos AP (2001) A statistical classification of Mediterranean species based on their flammability components. Intl J Wildland Fire 10(2):113–118. doi:10.1071/WF01004

Dimitrakopoulos AP (2002) Mediterranean fuel models and potential fire behaviour in Greece. Intl J Wildland Fire 11(2):127–130. doi:http://dx.doi.org/10.1071/WF02018

Eyre FHE (1980) Forest cover types of the United States and Canada. Soc Am For, Washington, DC, pp 88

Fahnestock GB (1970) Two keys for appraising forest fire fuels. USDA Forest Service, Pacific Northwest Forest and Range Experiment Station, Research Bulletin PNW-RB-099, Portland, pp 31

Fernandes P (2009) Combining forest structure data and fuel modelling to classify fire hazard in Portugal. Ann For Sci 66(4):415–415. doi:10.1051/forest/2009013

Finney MA (1998) FARSITE: Fire Area Simulator – model development and evaluation. United States Department of Agriculture, Forest Service Rocky Mountain Research Station, Research Paper RMRS-RP-4. Fort Collins, pp 47

Forestry Canada Fire Danager Group FCFDG (1992) Development and structure of the Canadian forest Fire Behavior Prediction System. Forestry Canada, Science Sustainable Development, Information Report ST-X-3, Ottawa, pp 62

Gauch HG, Jr, Whittaker RH (1981) Hierarchical classification of community data. J Ecol 69(2):537–557

Hawkes B, Niemann O, Goodenough D, Lawson B, Thomson A, Sahle W, Fuglem P, Beck J, Bell B, Symingto P (1995) Forest fire fuel type mapping using GIS and remote sensing in British Columbia. In: Heit M, Parker HD, Shortreid A (eds) Proceedings of the symposium GIS Applications in Natural Resources 2 – The 9th sympo/sium on Geographic Information Systems, Vancouver, pp 290–299

Hessburg PF, Reynolds KM, Keane RE, James KM, Salter RB (2007) Evaluating wildland fire danger and prioritizing vegetation and fuels treatments. For Ecol Manag 247:1–17

Hornby LG (1935) Fuel type mapping in Region One. J For 33(1):67–72

Hornby LG (1936) Fuel type mapping, its relation to forest fire control planning. Northwest Sci 2:3–7

Keane RE (2013) Describing wildland surface fuel loading for fire management: a review of approaches, methods and systems. Intl J Wildland Fire 22(1):51–62. doi:http://dx.doi.org/10.1071/WF11139

Keane RE, Burgan RE, Wagtendonk JV (2001) Mapping wildland fuels for fire management across multiple scales: Integrating remote sensing, GIS, and biophysical modeling. Intl J Wildland Fire 10:301–319

Keane RE, Frescino TL, Reeves MC, Long J (2006) Mapping wildland fuels across large regions for the LANDFIRE prototype project. In: Rollins MG, Frame C (eds) The landfire prototype project: nationally consistent and locally relevant geospatial data for wildland fire management. USDA Forest Service Rocky Mountain Research Station, General Technical Report RMRS-GTR-175, Ogden, pp 367–396

Keane RE, Reeves MC (2011) Use of expert knowledge to develop fuel maps for wildland fire management. In: Perera A, Drew A (eds) Expert knowledge and its application in landscape ecology. Island Press, New York, pp 211–228

Keane R, Gray K, Bacciu V, Leirfallom S (2012a) Spatial scaling of wildland fuels for six forest and rangeland ecosystems of the northern Rocky Mountains, USA. Landscape Ecol 27(8):1213–1234. doi:10.1007/s10980-012-9773–9

Keane RE, Gray K, Bacciu V (2012b) Spatial variability of wildland fuel characteristics in northern Rocky Mountain ecosystems. USDA Forest Service Rocky Mountain Research Station, Research Paper RMRS-RP-98, Fort Collins, pp 58

Keane RE, Gray K (2013) Comparing three sampling techniques for estimating fine woody down dead biomass. Intl J Wildland Fire 22(8):1093–1107. doi:http://dx.doi.org/10.1071/WF13038

Keane RE, Herynk JM, Toney C, Urbanski SP, Lutes DC, Ottmar RD (2013) Evaluating the performance and mapping of three fuel classification systems using Forest Inventory and Analysis surface fuel measurements. For Ecol Manag 305:248–263. doi:http://dx.doi.org/10.1016/j.foreco.2013.06.001

Keyes CR (2002) Quantifying stand targets for silvicultural prevention of crown fires. Western J Appl For 17(2):101–109

Krivtsov V, Vigy O, Legg C, Curt T, Rigolot E, Lecomte I, Jappiot M, Lampin-Maillet C, Fernandes P, Pezzatti GB (2009) Fuel modelling in terrestrial ecosystems: an overview in the context of the development of an object-orientated database for wild fire analysis. Ecol Model 220(21):2915–2926

Lutes DC, Keane RE, Caratti JF (2009) A surface fuels classification for estimating fire effects. Int J Wildland Fire 18:802–814

McKenzie D, Raymond CL, Kellogg L-KB, Norheim RA, Andreu A, Bayard AC, Kopper KE, Elman E (2007) Mapping fuels at multiple scales: landscape application of the Fuel Characteristic Classification System. Can J For Res 37:2421–2437

Muraro SJ (1965) Fuel classification – a review of the literature and discussion of principles. Paper presented at the Second federal fire research conference, Victoria, British Columbia, Canada, March 22–26

Orloci L (1967) An agglomerative method for classification of plant communities. J Ecol 55(1):193–206

Ottmar RD, Sandberg DV, Riccardi CL, Prichard SJ (2007) An overview of the Fuel Characteristic Classification System – Quantifying, classifying, and creating fuelbeds for resource planning. Can J For Res 37:2383–2393

Parsons RA, Mell WE, McCauley P (2010) Linking 3D spatial models of fuels and fire: Effects of spatial heterogeneity on fire behavior. Ecol Model 222(3):679–691

Poulos HM, Camp AE, Gatewood RG, Loomis L (2007) A hierarchical approach for scaling forest inventory and fuels data from local to landscape scales in the Davis Mountains, Texas, USA. For Ecol Manag 244(1–3):1–15

Reeves MC, Kost JR, Ryan KC (2006) Fuels products of the LANDFIRE project. In: Andrews PL, Butler BW (eds) Fuels management – how to measure success, Portland OR. USDA Forest Service Rocky Mountain Research Station, Proceedings RMRS-P-41, Fort Collins, pp 239–249

Reich RM, Lundquist JE, Bravo VA (2004) Spatial models for estimating fuel loads in the Black Hills, South Dakota, USA. Intl J Wildland Fire 13:119–129

Reinhardt E, R.E. Keane, Brown JK (1997) First Order Fire Effects Model: FOFEM 4.0 User's Guide. USDA Forest Service, Intermountain Research Station General Technical Report INT-GTR-344, Ogden, pp 65

Reinhardt ED, Keane RE, Brown JK (2001) Modeling fire effects. Intl J Wildland Fire 10:373–380

Riccardi CL, Ottmar RD, Sandberg DV, Andreu A, Elman E, Kopper K, Long J (2007a) The fuelbed: a key element of the fuel characteristic classification system. Can J For Res 37:2394–2412

Riccardi CL, Prichard SJ, Sandberg DV, Ottmar RD (2007b) Quantifying physical characteristics of wildland fuels using the fuel characteristic classification System. Can J For Res 37:2413–2420

Rothermel RC (1972) A mathematical model for predicting fire spread in wildland fuels. United States Department of Agriculture, Forest Service, Intermountain Forest and Range Experiment Station, Ogden, Utah Research Paper INT-115, pp 88

Sandberg DV, Ottmar RD, Cushon GH (2001) Characterizing fuels in the 21st century. Intl J Wildland Fire 10:381–387

Sandberg DV, Riccardi CL, Schaaf MD (2007) Reformulation of Rothermel's wildland fire behaviour model for heterogeneous fuelbeds. Can J For Res 37(12):2438–2455

Santoni P-A, Filippi J-B, Balbi J-H, Bosseur F (2011) Wildland fire behaviour case studies and fuel models for landscape-scale fire modeling. J Combust 2011:1–12. doi:10.1155/2011/613424

Scott J, Burgan RE (2005) A new set of standard fire behavior fuel models for use with Rothermel's surface fire spread model. USDA Forest Service Rocky Mountain Research Station, General Technical Report RMRS-GTR-153, Fort Collins, pp 66

Shiflet TNE (1994) Rangeland cover types of the United States. Society of range management, Denver, pp 123

Sikkink P, Keane RE, Lutes DC (2009) Field guide for identifying fuel loading models. USDA Forest Service Rocky Mountain Research Station, General Technical Report RMRS-GTR-225, Fort Collins, pp 44

Weise DR, Wright CS (2014) Wildland fire emissions, carbon and climate: Characterizing wildland fuels. For Ecol Manag 317(0):26–40 doi: http://dx.doi.org/10.1016/j.foreco.2013.02.037

Chapter 8
Fuel Sampling

Not everything that can be counted counts, and not everything that counts can be counted
Albert Einstein

8.1 Background

Directly measuring fuel properties in the field is the most accurate and consistent method for fire managers and scientists to collect the inputs needed for fuel description and fire behavior and effects simulation, especially when compared to fuel classification approaches (Chap. 7). Quantification of these properties is generally accomplished by *field sampling*; measuring fuel characteristics *in situ* to estimate fuel properties. And since there is a great diversity of fuel components (Chap. 3), coupled with a large number of fuel characteristics (Chap. 2), there are numerous sampling designs to estimate fuel properties at the particle, component, and fuelbed scale. Here, field sampling is a general term used to describe the wide range of approaches for measuring fuel properties for fuel components, including designing sample projects, conducting measurements in the field, and creating databases from measured information. Sampling design is easily the most important aspect of field sampling and it includes deciding on the sampling intensity (e.g., number plots), sample locations (e.g., random, stratified random), methods (e.g., planar intercept, fixed area plots), protocols (e.g., size classes, plot sizes), and techniques (e.g., caliper measurements). And since the majority of operational and research fuel sampling methods were designed to estimate loading, this chapter will focus on those sampling methods that estimate loading for a variety of surface fuel components. Methods, protocols, and techniques for estimating other fuel properties, such as mineral contents and particle densities, were mostly developed for research so they are difficult and costly to employ under operational sampling efforts and therefore, are not discussed here. This chapter also includes a general description of the common methods used to measure canopy fuel characteristics in the field.

R. E. Keane, *Wildland Fuel Fundamentals and Applications*,
DOI 10.1007/978-3-319-09015-3_8

Sampling designs can be stratified into two broad categories—inventory and monitoring. Inventory techniques are used to quantify fuel characteristics for one point in time, usually for planning and designing fuel treatments and describing fuel hazard and fire risk (Sampson and Sampson 2005). Monitoring involves sampling to estimate current status and detect change over time, such as that resulting from fuel treatments or fuel accumulation. There are usually two or more measurements at the same exact place but at different times (e.g., pretreatment, 1 year post-treatment, and at 5 year intervals after treatment). Monitoring protocols are often quite rigorous because they must detect subtle changes over time, sometimes across multiple spatial scales, so more plots and more detailed measurements are sometimes necessary. Both inventory and monitoring are critical tasks for fire management, yet there are few standardized efforts to collect field data on fuel properties across nearly all fire management agencies. To implement effective fuel treatment programs, which cost millions of dollars, and to assess the efficacy of these treatments, a comprehensive, standardized fuel sampling program is a critical tool to facilitate enlightened and adaptive fire and fuel management.

Inventory and monitoring fuel sampling methods are often designed for one of two broad objectives—research or management. Research sampling techniques are usually uniquely specialized to quantify some set of fuel properties with high precision and accuracy to satisfy research objectives. They are often tailored to answer a specific research question, and because of this, these methods are usually quite intense (large samples, many plots), highly localized, time-consuming, costly, and often requiringe highly skilled personnel and specialized equipment. Since most research sampling techniques were designed around specific study objectives and study areas, they are often difficult to apply in broader situations, such as operational fire management. There are many physical fuel properties measured in research studies, such as heat content and specific gravity, which change little across fuel particles relative to particle abundance, so these research results are often used IN many management applications (Nalder et al. 1999).

Management-oriented or operational sampling is often done to facilitate the planning, design, and eventual implementation of a fire management project. Often, these sampling designs do not require the same degree of accuracy as research sampling, so they are often less intensive, not as costly, and easier to implement (Lutes et al. 2006). Management sampling efforts are often designed to be applied across large areas by field technicians with little to high levels of training in fuel sampling. Sampling techniques designed for managers are also often highly generalized so that they can be applied across diverse areas and situations. This may result in the application of sampling techniques or protocols that may be inappropriate for a particular ecosystem or treatment area. Logs, for example, may be sampled using an insufficient number of transects to meet a desired level of accuracy, especially in those ecosystems where logs are scarce (Sikkink and Keane 2008). The main topic of this chapter is how to sample fuel biomass for each fuel component. However, knowledge of the basics of fuel sampling is first needed to fully understand the use of the methods presented here.

8.2 Sampling Basics

The first and most important step in any fuel sampling effort is to clearly articulate a sampling objective. Many fuel sampling efforts were unsuccessful because there never was a comprehensive statement of the purpose for the sampling. A well-written sampling objective will guide all other decisions involved in designing sampling projects (Lutes et al. 2006). Sampling objectives should follow the S.M.A.R.T. guidelines: Specific, Measureable, Achievable, Relevant, and Time-based (Lutes et al. 2006). Many people make the mistakes of (1) using goals instead of defining objectives (not specific), (2) failing to mention what is being sampled (hard to measure), (3) specifying too many objectives for a sampling effort (not achievable), (4) including aspects that are unrelated to the sampling effort (irrelevant), and (5) forgetting to add deadlines and scheduling concerns (not time-based). Without doubt, a well-stated objective is the keystone of a successful sampling design.

Many fuel loading sampling efforts estimate loadings for a defined area (target population) using a set of opportunistically, randomly, or systematically located sampling units. Sampling units, such as fixed-area plots (FAPs) or planar intercept (PI) transects, are used to sample the target area. Measurements taken within the sampling units that are distributed across the sample area are then summarized to compute an estimate of loading for the sample area. The first step in fuel sampling is delineating the sample area and then deciding on a sampling unit (Catchpole and Wheeler 1992). Sampling unit selection is important because it will dictate the logistics of sampling design. The sample unit can be FAPs, transects, planes, or points. FAP sampling units, for example, can vary in size from macroplots (generally 100–1000 m^2) to microplots (~1–100 m^2) to nanoplots (~0.1–1 m^2) to dimensionless points, depending on the fuel component sampled, available resources, and the stated objective. Each sampling effort employs a unique design developed specifically to meet sampling objectives while considering important sampling constraints, such as time, cost, and available expertise.

All ecological sampling projects, but especially wildland fuel loading sampling efforts, are designed using a delicate balance between ecological, logistical, and resource concerns. The most important factor in operational management fuel inventory and monitoring projects is the amount of resources available for sampling. In the end, most field fuel sampling efforts reflect a compromise in some resource limitation. The most important resource is funding because with adequate funding, most of the other resource limitations can be mitigated (e.g., hire more people, buy more equipment). The next valuable resource is time. Many sampling efforts were unsuccessful because it was impossible to both collect and report the critical data in the time frame allowed. The number of qualified people available to assist in the sampling effort may also dictate sampling designs. And last, transportation, safety, and equipment resources are also important to consider in sampling designs. Vehicles to get crews to remote locations on rough gravel roads may be a limiting factor, as is the lack of sampling equipment required for fuel measurement, including the critical gear that ensures crew safety such as radios, first aid kits, and cell phones.

Expertise of the field crew and the people that will ultimately analyze and interpret the data is important in fuel sampling. Inexperienced field crews will require intensive training that may reduce the time available for sampling. And similarly, inexperienced analysts may produce inappropriate statistical summaries and come to the wrong conclusions, while inexperienced managers may use completed analysis results in inappropriate contexts that don't fit sampling objectives or sampling designs. Fuel sampling personnel may be highly experienced, who can easily adapt to any challenge in the field without significant changes in productivity and quality, to novice student summer temporary hires, who have difficulty navigating in the field let alone accurately measure fuel characteristics. Effective training is the only remedy for inexperienced sampling crews.

The last important factor is the level of statistical rigor demanded by the sampling project, which should always be determined in the context of the sampling objective. One of the most important parameters in the sample design is the number of sample units (n) to establish in the sample area to obtain a statistically credible loading estimate, often called the sampling intensity. This is done using the following formula:

$$n = \left[\frac{z\sigma}{E}\right]^2, \quad (8.1)$$

where E is the difference between the sampled mean value (i.e., loading) and the population mean loading value, σ is the population variance, and z is the z value for tail of the t distribution for a selected probability value α often selected as $\alpha = 0.05$ for most sampling projects. E is estimated by how close the sampler wants to be to the population mean (e.g., 20 % of the population mean). To calculate n, most sampling projects need an *a priori* (beforehand) estimation of the loading variability (σ) and the population mean to compute the number of sampling units needed for a statistically credible estimate. The problem is that the statistical parameters (E, σ) for fuel loading depend on the fuel component, and the variability of each fuel component loading is highly localized and is different by region, ecosystem, topographic setting, and time since disturbance (Keane et al. 2012a). Therefore, a priori population mean variabilities by component are difficult to estimate from past projects. Moreover, the factors mentioned above (resources, expertise) may often overwhelm requirements for statistical rigor in some sampling projects. Requiring an unachievable number of sample units given resource limitations to satisfy a statistical requirement may be counterproductive. Likewise, executing a sampling program that cannot hope to address the project's objectives because of inadequate precision also makes poor use of available resources. Statistical rigor must be balanced with the other factors to construct a successful sampling design.

In summary, there are usually several tasks that must be done to design an effective sampling projects: (1) identify the number of people available and assess their expertise, (2) estimate the time allowed to conduct the sampling, (3) delineate and describe the sample area(s) that need sampling, (4) select an appropriate sampling unit to get the project done on time and with the people and resources available, and (5) decide on the equipment required for sampling. This information can then be used to design a sampling approach using the following example. Assume there is a 100-ac treatment unit that will be monitored for changes in fuel loading and potential smoke emissions and there is only 1 month (20 working days) to accomplish the initial measurements of the monitoring project. By selecting the First Order Fire Effects Model (FOFEM) program to estimate emissions (Reinhardt et al. 1997), the set of surface fuel components that need loading estimates are identified as litter, duff, 1, 10, 100, 1000 h, shrub, and herb (Table 3.1). A 0.1 ac circular macroplot is selected as the sampling unit because protocols for measuring all eight surface components and canopy fuels are easily nested in this FAP. Assuming a two-person crew and a 1-h sampling time to measure loading for all eight components including travel to the next plot, and assuming 1 h in each 8-h day is used for transportation to and from the site, we can then estimate the potential number of plots for this project as 140 (seven plots per day, 20 working days). This means that approximately 14 % of the project area may be sampled for those fuel components (140 plots × 0.1 ac = 14 ac of 100 ac). Estimates of variability for each fuel components obtained from other projects or from the literature (see Keane et al. 2012b) can be used to compute the number of plots needed for a statistically credible sample (see Eq. (8.1) or Lutes et al. 2006). The statistical estimate can be compared with the 140 plots to adjust the design criteria to create a successful sample design, such as adding more people, increasing sampling time (work 10 h days, add 10 days), reducing number of fuel components measured, or modifying the sample design (e.g., distribute plots systematically or among strata). The last important sampling item to be selected is the surface or canopy fuel sampling technique to use at each of the plots. The next two sections detail the diverse methods often employed by fuel specialists to sample fuel characteristics.

8.3 Surface Fuel Loading Sampling

Numerous techniques and methods have been developed to estimate surface fuel loading for both research and management to allow for greater flexibility in matching available resources with sampling objectives and constraints (Catchpole and Wheeler 1992). These techniques are arranged below in order from easiest to most difficult with a corresponding gradient from most to least uncertain (Table 8.1). The first set of indirect methods is not recommended, but many fire managers have used these methods in the past to estimate fuel loadings when no other information is available and there aren't resources for other alternatives.

Table 8.1 Comparison of possible surface fuel sampling methods and techniques to estimate fuel loading for forest and rangelands. In general, the techniques are arranged in the order of easiest with most variability and uncertainty to the most difficult but more accurate and precise. Abbreviations and codes are as follows: Fuel component are the fuel components in Table 3.1 that can be sampled with this method; sampling time is the time it takes to get estimates (1: fast, 10: slow), level of uncertainty is the amount of uncertainty in fuel loading estimates (1: highly certain to 10: highly uncertain), cost is the relative cost of implementing technique (1: inexpensive to 10: costly), and training needed is how much training is needed to get acceptable estimates of fuel attributes (1: low to 10: high)

Technique	Description	Advantages	Disadvantages	Fuel components	Sampling time	Level of uncertainty	Cost	Training needed
Indirect methods—association techniques								
Existing data	Use existing data from other areas to extrapolate to the project area	Many fuel attributes may be available	Fuel conditions are highly variable across sites depending on disturbance history	All	1	8	3	2
Vegetation	Existing data are summarized by vegetation type and these summaries are assigned to project area based on vegetation type	Most people can identify vegetation types	Vegetation not related to fuels	All	3	7	4	2
Map value	Use mapped values as fuel estimates	Represents local area	Fuel maps have low accuracies	All	1	9	2	1
Indirect methods—visual techniques								
Eye estimate	Estimating loading from visual inspection	Integrates local knowledge	Highly inaccurate with high user bias. Requires extensive training	All	3	7	3	8
Photo series	Use a local photo series publication to visually assess fuel loadings at the project area	Many series available; quantify many fuel comp. and attributes	Requires training; rarely assessed for accuracy	All	4	6	3	6
Photoload	Use a series of photographs that show fuelbeds with gradually increasing loading	One set of photos good for most woody fuelbeds	Doesn't include all fuel components; requires training	1, 10, 100, and 1000 h, shrub, herb	5	5	3	7

Table 8.1 (continued)

Technique	Description	Advantages	Disadvantages	Fuel components	Sampling time	Level of uncertainty	Cost	Training needed
Fuel classification	Use a field-keyed fuel classification category to obtain the fuel estimates	Direct link to fuel map entities and sampling estimates	Few classifications key fuelbed characteristics	All	4	6	3	5
Fuel hazard assessments	Rate hazard for various fuel layers and components on an ordinal scale and correlate to loadings	Links hazard assessment to fuel loadings; easy to teach and use	Must be developed for each ecosystem; specific to Australian fuel components	Litter, Fine Woody Debris (FWD), shrub, herb	5	5	3	6
Visual cover-volume	Use canopy cover and height to estimate volume, and then use volume and bulk density to estimate loading for some fuel components	Often the only efficient technique for measuring shrub, herb, duff, and litter loadings	High variability in cover estimates coupled with high variability in height (depth) measurements	Litter, duff, shrub, herb	4	5	3	5
Indirect methods—imagery techniques								
Image processing	Analyze imagery to measure fuel conditions	Visual record of fuelbed	Imagery not correlated to some fuel attributes	Depends on imagery	4	5	8	9
Terrestrial LiDAR	Use scanning LiDAR to obtain a profile of the fuelbed to estimate fuel attributes	Locations of all fuel particles are measured	Can't relate LiDAR strikes to material that reflected the strike	Depends on the analysis	7	4	10	9
Direct methods								
Planar intercepts	Estimate fuel attributes along a vertically oriented sampling plane	Readily available and standardized methods available	Only useful for woody fuels	1, 10, 100, 1000 h	7	7	7	6
Fixed-area microplots	Estimate fuel attributes within a plot frame	Adjust plot size to fuel component distributions	High variability across microplots; may require many microplots	All	8	8	8	7

Table 8.1 (continued)

Technique	Description	Advantages	Disadvantages	Fuel components	Sampling time	Level of uncertainty	Cost	Training needed
Perpendicular distance sample (PDS)	Use PDS methods to estimate fuel loading by measuring distances or using a prism or other similar device	Relatively quick and easy	Training required; only used for one component	Coarse woody debris (CWD)	3	7	5	8
Measured cover-volume	Measure cover using standard ecological sampling protocols, then measure component heights to get volume and multiply by bulk density to get loading	Useful for novel fuelbeds, best method for litter and duff	Not useful for woody fuels, cover poorly correlated to fuels	Litter, duff, shrub, herb	7	4	6	5
Destructive	Collect fuel within a defined area, bring it back to lab and dry it, then weigh it to get loading	Highest accuracy in measuring loadings	Too costly and time-consuming for operational applications	All	9	2	9	4

8.3.1 *Indirect Methods*

These methods involve quantifying fuel loadings using techniques that do not directly involve measuring the fuel property, but rather, use other references or sources to quantify loadings. This usually involves subjectively assigning loadings by comparing with existing data (association), inspecting fuel conditions and visually comparing to reference conditions (visual), or correlating with remotely sensed imagery (imagery).

8.3.1.1 Associative Techniques

The most common associative technique involves using *existing data* or information, often collected by someone else from somewhere else, to estimate loading values for the area of concern or project area. Fuel loading data collected for another area, for example, may be assigned to the area in question if the two areas are deemed similar, perhaps based on vegetation composition, disturbance histories, and biophysical site conditions. Catchpole and Wheeler (1992) call this approach the comparative yield method and mention that they could be improved by using statistics, photos, and expertise to aid in the data assignment. The problem with this technique is that each site and project area is ecologically unique and the extrapolation of loadings from one site to another might ignore those important but subtle factors that have influenced component loadings, such as differences in basal area, tree density, disturbance history, topographic setting, and stand structure.

Another commonly used associative technique is to assign fuel loadings to a sample area based on the sampled area's *vegetation* characteristics, similar to association approach used in fuel classification (Chap. 7), except, in this case, it is used to assess actual loadings in the field. Several fuel classifications were built by summarizing plot-based fuel component loadings across categories in vegetation and related classifications such as structural stage, cover type, and potential vegetation type. The FOFEM program, for example, includes loading defaults as a summary of fuel loadings across legacy plots within SAF (Society of American Foresters) and SRM (Society of Range Management) cover type categories (Reinhardt et al. 1997). Many people have used these defaults as a fuels inventory when conducting various analyses. The Fuel Characteristics Classification System (FCCS) (Ottmar et al. 2007) was specifically designed so that fuel loading data collected for one area could be used for other areas based on a set of seven vegetation and disturbance-related stratifications. This indirect method assumes that fuel component loading values, either individually or as a collective group, correlate to vegetation characteristics. However, as in fuel classification development (Chap. 7), studies have found that fuel loadings correlate poorly to vegetation types, especially at the fine spatial scales of project and treatment areas (Brown and Bevins 1986; Keane et al. 2012b, 2013). Vegetation classification categories may correlate to fuels at coarser scales (e.g., lifeforms, large regions), but the high variability of fuels at fine scales often overwhelms differences across broad vegetation types (Keane et al. 2012b). Disturbance history (e.g., time since disturbance, severity) is more important than vegetation to predict fuel loadings (Brown and Bevins 1986), but few studies have explored this relationship. Some have found that a variation of this vegetation associative technique is useful to create predictive loading equations from measured vegetation characteristics using statistical methods (Catchpole and Wheeler 1992). Fuel loadings can be correlated to various stand-related characteristics, such as basal area, Leaf Area Index (LAI), and stand height. Keane et al. (2012b) found that stand attributes were poorly correlated to surface fuel loadings but were highly correlated to canopy fuel variables.

Another associative method is using mapped loading values from readily available *digital geospatial products* as fuel loading estimates (Chap. 9). The LANDFIRE National Project, for example, mapped four fuel classifications across the United States using satellite imagery (Reeves et al. 2006) and many have used the loading values from these classifications to quantify loadings for a specific project area. However, Keane et al. (2013) found low accuracies for fuel loadings from LANDFIRE fuel maps. Therefore, this practice, while inexpensive, quick, and easy, is not recommended for fine scale, project-level applications until existing fuel maps are much improved. Locally created fuel maps may have sufficient quality, but regional and national maps should only be used for fuel analyses at broad scales, not at the project level. Depending on the resolution, fuel maps could still be useful for stratifying the sample area (or target population) into more homogeneous subunits to make sampling efforts more efficient.

8.3.1.2 Visual Techniques

Visual techniques involve assessing the loading of fuel components from ocular estimates. Some fuel specialists feel they can accurately estimate loadings by *eye* without any guides or references. This level of resolution and accuracy may be acceptable for some fuel applications, such as describing fuels to other professionals. However, it is rare that anyone can accurately and consistently estimate the loadings of all fuel components by eye, especially for FWD, duff, and litter. One reason for this is that fuel loadings have high spatial variability over small areas (Chap. 6) and the ocular estimate is often biased toward smaller portions of the project area; it is difficult to evaluate a large, heterogeneous area to obtain a truly integrated visual fuel loading estimate (Sikkink and Keane 2008). While visual estimation by eye is preferable to some of the associative methods presented above providing there is a high expertise and confidence in the sampler, it is rare that a person can accurately estimate fuel component loadings across diverse fuelbeds at the same detail. Therefore, many have resorted to using pictures as guides and references for comparing loading estimations.

Perhaps the most popular comparative visual technique is the *photo series,* which is both a classification (Chap. 7) and a fuel assessment technique. Using photo series sampling methods, surface fuel loadings are ocularly estimated using a set of photos that present stand conditions for various vegetation types and site conditions (Fig. 7.1). Photos were taken of representative fuel types in a particular geographical region, and then fuel component loadings were measured for the photo footprint and the summary of those loadings is reported next to the photo in the photo series publication. These photo series publications are taken to the field and the observed conditions in the field are visually matched to the best photo and the loading measured for the photographed stand are used for the loadings of the matched stand. A different photo can be used to estimate each of the various fuel components. Often, photo series fuel types are stratified by vegetation conditions (cover type, structural stage).

The photo series was introduced by Maxwell (1976), improved upon by Koski and Fischer (1979) and Fischer (1981), and then extrapolated across the USA (http://www.fs.fed.us/pnw/fera/research/fuels/photo_series/). Many have taken the photo series concept and applied it to areas that have been treated (Koski and Fischer 1979), experienced severe disturbances (Vihnanek et al. 2009), contain special vegetation types (Ottmar and Vihnanek 2000), and found in other countries (Morfin-Ríos et al. 2007). Others have adapted the photo series concept to use three-dimensional stereoscopic photos (Vihnanek et al. 2009). Photo series data comprise the majority of the national fuelbeds in the FCCS database that have been mapped across the USA by McKenzie et al. (2007) and Reeves et al. (2009). A list of completed photo series for the US Rocky Mountains is presented in Baker (2009).

And most importantly, photo series have been developed for many local settings to be applied at fine scales within a small geographical region. One highly valuable aspect of the photo series is that the fire behavior fuel model (FBFM, Chap. 7) is often documented along with fuel loadings so that FBFMs can be more easily assessed in the field.

Despite its huge popularity, the photo series sampling technique has yet to be comprehensively evaluated across many vegetation types or environmental conditions. Sikkink and Keane (2008) found loading estimated using photo series approaches were often inaccurate and difficult to repeat across observers, albeit there were some limitations in the training of the crews. Many photo series photos emphasize stand-level differences with oblique photos, and, as a result, some fine fuel components, such as FWD, litter, and duff, may be hidden by the vegetation in the photo or are undetectable because of their small size. Additionally, the photo series cannot be used to assess the loading of duff or litter because the photos do not show their profile depth. While photo series may give loading estimates to the resolution needed for management decisions, other uses of loading estimates, such as predicting smoke emissions and carbon inventories, may demand a more accurate and repeatable method of loading estimate.

A new method of visually assessing fuel loading has been developed to improve on photo series techniques and to compete with other direct sampling methods (Sect. 8.3.3). The *photoload* method uses calibrated, downward-looking photographs of known fuel loads for woody, shrub, and herbaceous fuels to compare with conditions in the field (Keane and Dickinson 2007a, b). These ocular estimates can then be adjusted for diameter, rot level, and fuelbed height. There are different photoload methods for logs, FWD, shrubs, and herbaceous material, but there are no photoload methods for measuring duff and litter loading. The photoload technique differs from photo series in that assessments are made by comparing field fuel conditions to smaller-scale downward-pointing photographs of graduated fuel loadings. Photoload methods are much faster and easier than more complicated techniques (Sect. 8.3.2) with comparable accuracies (Sikkink and Keane 2008), and they can be used in multistage sampling strategies where a fraction of the total plots are also destructively sampled and correlated to photoload samples to develop a means for correcting all photoload estimates (Keane et al. 2012b). However, Keane and Gray (2013) found the photoload technique requires extensive training to be used effectively; inexperienced users often could not consistently and accurately estimate high fuel loads.

Fuel classifications can also be used as an inventory and monitoring method (Chap. 7). In this technique, a fuel classification class is visually identified in the field, and the loadings assigned for that class are used as the sampled loadings. Those fuel classifications that use vegetation to classify fuelbeds are probably the most suspect, while classifications that contain dichotomous keys for identifying classes based on fuelbed properties, such as the Fuel Loading Model (FLM) classification (Lutes et al. 2009b), are best for fuel assessment because they can be used in the field by inexperienced crews to estimate fuel loadings with moderate accuracies (Keane et al. 2013).

Another effective visual fuel sampling method uses fuel hazard assessments across different fuel strata to obtain loading estimates for various components in the fuelbed. Originally developed by Gould et al. (2008) for Australia, this method involves making hazard assessments for the overstory and intermediate canopy layers, and then elevated, high, and low surface fuel layers. Each layer is given a score from 0 to 4 based on a variety of fuelbed attributes including percent canopy cover, presence of stringy bark, and suspended dead material. These scores are then summarized and the summaries are correlated to actual fuel loadings using statistical techniques (Gould et al. 2011). This rapid technique produced moderately accurate loadings with minimal training. Techniques that successfully link visually distinctive signatures, such as canopy cover, with fuel component loadings might be effective for operational fuel sampling in the future because it balances and integrates the elements of hazard assessment into the sampling design.

One last visual technique involves using *cover-volume methods* to calculate loadings from visually estimated canopy cover and height. In this technique, canopy cover is estimated by eye for those components with small and variable fuel particles that are grouped together into one component, such as shrubs, herbs, and trees, and an estimate of measured or ocularly estimated height is also made in a fixed-area sample unit for those components. Some fuel sampling packages, such as FIREMON (Lutes et al. 2006), describe how to estimate canopy cover in 10% classes (e.g., 1–5, 5–15, 15–25%, and so on) and how to visually estimate height. Volumes of the assessed components (volume includes air pockets) are then calculated by multiplying the proportion cover (percentage cover divided by 100) by height (m) and sampling area (m^2). Fuel loadings are then estimated by multiplying volume (m^{-3}) by bulk density estimates ($kg\ m^{-3}$) for the sample unit. Bulk densities for litter, duff, shrub, and herb components can be found in the literature (Brown 1981; Keane et al. 2012b) or destructively sampled for a small proportion of the plots. This is often the only operational method for estimating fuel loadings for these complex fuel components. Sneeuwjagt (1973) used a variation of this approach when he developed equations that predicted loading from height for both litter and shrubs. While canopy cover is used extensively in plant ecology studies (Mueller-Dombois and Ellenberg 1974), it requires extensive training to visually estimate cover of multiple, overlapping fuel components to the accuracy and precision needed for fuel sampling.

8.3.1.3 Imagery Techniques

Imagery techniques involve using advanced statistical analysis to correlate fuel loadings to the digital signatures in the digital imagery. This imagery is taken in the field and is different from the airborne or satellite imagery used for fuel mapping (Chap. 9). A potentially useful imagery technique is the quantification of fuel loads using *image processing* techniques or software. Many years ago, Fahnestock (1971) calculated loading for several fuel components using a dot grid projected on color photographs of a cross-section of bayberry shrub fuel layer. Today, there are more sophisticated image processing approaches that use computer software. The stereoscopic vision technique (SVT), for example, involves taking stereoscopic photos of the fuelbed in the field then inputting the digital photos into computer-image recognition software to identify woody fuels and then compute loading volume (Arcos et al. 1998; Sandberg et al. 2001). Others have taken pictures of the fuelbed and then attempted to quantify loading using advanced image processing techniques (Jin 2004). Photographic methods are still under development and there needs to be major gains in image processing to discriminate between fuel components and compute volumes. Its primary use is in quantifying CWD loading (Arcos et al. 1998), but it may find some eventual use for measuring FWD, shrub, and grass loading.

Another emerging technology is the use of *ground-based LiDAR* to estimate fuel loads for some fuelbeds (Loudermilk et al. 2009). Here, a terrestrial scanning LiDAR (TSL) unit is mounted on a truck or some other vehicle to obtain scan distances for ground fuels at subcentimeter scales. The LiDAR signal can then be related to loading by constructing statistical models where destructively sampled loadings for various components are correlated to the LiDAR imagery scan distances. It is sometimes difficult to differentiate between fuel components using TSL in heterogeneous fuelbeds, but it is still possible. This technique may only be possible for research purposes in the near future because the TSL instrument is rather expensive (>$ 40,000), demands a high level of expertise to use and analyze, and it is also difficult to transport and use in complex terrain.

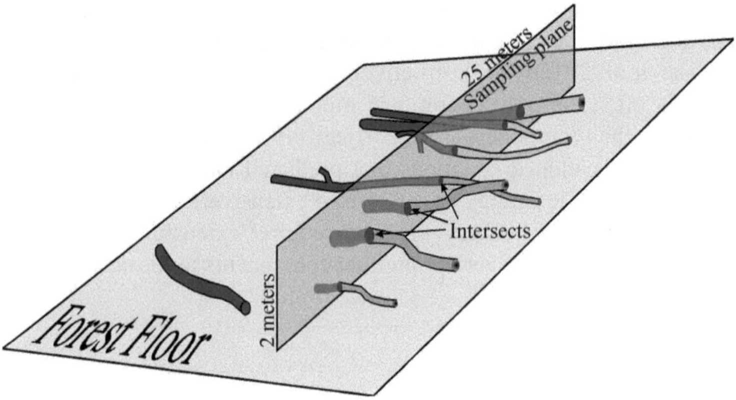

Fig. 8.1 Illustration showing the sampling of fuel particles using the planar intercept method

8.3.2 Direct Methods

These methods involve directly sampling or measuring characteristics of fuel particles to calculate loading. This usually involves direct contact with the fuel, such as measuring dimensions of particles using calipers, estimating depths of duff and litter using rulers, or collecting particles for drying and weighing in the lab.

8.3.2.1 Planar Intercept (PI)

PI techniques are the most commonly used sampling methods for sampling downed woody fuels for both management and research (Catchpole and Wheeler 1992; Dibble and Rees 2005) and both inventory and monitoring projects (Busing et al. 2000; Waddell 2001). PI sampling involves counting woody fuel particles by diameter size classes, or by directly measuring individual particle diameters, as they intercept a vertical sampling plane that is of a fixed length and height (Brown 1970, 1974; Fig. 8.1). These intercepts are then converted to loadings using standard formulae (Brown 1974). There are correction factors for the slope of the sampling plane and orientation of the fuel particles. One major parameter for computing loading is the particle density (specific gravity) of the particles (Table 2.1; Chap. 2).

PI is the operational version of the line transect method originally introduced by Warren and Olsen (1964) and made applicable for measuring CWD by Van Wagner (1968). The line transect method is founded in the probability-proportional-to-size concepts and several variations of it have been developed since 1968, including those that vary the transect arrangements and directions, and those that apply the technique using different technologies (Hansen 1985; Nemec-Linnell and Davis 2002; Vries 1974). Brown (1971) modified the line transect method so that particle intercepts are measured in a two-dimensional plane rather than a one-dimensional line for operational sampling of FWD and CWD in forests (Brown 1974; Brown et al. 1982). Many still locate the bottom of the plane using a line transect that is often represented by a cloth tape.

The advantages of the PI method are that it is easy to use and easy to teach (Lutes et al. 2006, 2009a). Novice field technicians can be taught this method in a short time (1 h) to achieve moderately repeatable measurements. The method can also be easily modified to adjust for local conditions, available expertise, and sampling conflicts, such as long plot times, scattered woody fuels, and slash. The sampling plane can be any size, shape, or orientation in space and samples can be taken anywhere within the limits set for the plane (Brown 1971). It also requires few specialized equipment; often a plastic ruler and cloth tape are the only gear needed.

However, there are some problems to the PI method. First, it only can be used for estimating downed dead woody loading; loadings for other fuel components, such as canopy fuels, litter, and duff, must be estimated with entirely different methods. This is somewhat problematic because the sampling unit for PI (transect) does not always scale to the FAP methods used for sampling other components or used in other forest and range inventories (Keane and Gray 2013). The CWD transects, for example, are usually too long to fit within the area of standard fixed area plots. PI sampling designs are also difficult to merge with other sampling designs because the PI was designed to sample entire stands, not FAPs. PI methods also require a large number of transects under highly variable fuel conditions, which may be time- and cost-prohibitive for operational sampling efforts. Keane and Gray (2013) found that over 200 m of transect were needed on a 0.05 ha plot to sample FWD within 20% of the mean. Moreover, some feel that it is difficult to repeat particle intercept counts with any degree of reliability (Sikkink and Keane 2008). Particles are often hidden by other fuels often partially buried in the litter and duff making repeatability across and within observers difficult.

8.3.2.2 Fixed-Area Plots (FAP)

In contrast to unequal probability strategies (e.g., PI), FAP are based on equal probability sampling methods and have been adapted from vegetation composition and structure studies to sample fuels (Mueller-Dombois and Ellenberg 1974). In FAP sampling, a plot of any geometric shape, often round or square, is used as a sampling unit and all fuels within the plot boundary are measured using any number of fuel measurement methods including destructive collection (cut, dry, and weigh fuels; see Sect. 8.3.3.5), volumetric measurements (measure diameter, length to compute volume, then use density to estimate weight), vertical depths of duff and litter layers (measure thickness of duff and litter layer), and particle counts by size class (count particles, assume standard length, diameter, then compute weight) (Keane et al. 2012b). FAPs can be any size, and the most effective sampling efforts scale the size of the FAP to the fuel being measured. Because FAP approaches require significant investments of time and money, they are more commonly used to answer research questions rather than to monitor or inventory fuels for management planning. However, new methods have been designed to use FAP in operational sampling projects (Keane and Gray 2013).

The FAP method may be a preferable and more appropriate method for obtaining accurate fuel loading estimates for many surface fuel components. Most importantly, FAP techniques tend to give a better representation of the actual variation observed in the field for surface fuel components (Keane and Gray 2013). FAP sizes and number can be adjusted to reduce sampling times but may result in reduced precision of fuel loading estimates. FAP size can also be adjusted to account for the spatial scaling of loading by fuel size. Larger fuels (CWD), for example, can be sampled with larger plots to fully account for spatial distributions in sample estimates. Moreover, FAP sampling is easily adapted or merged with other protocols that are commonly used to sample other fuel components or other ecosystem attributes. Microplots used to sample FWD, for example, can also be used to sample tree seedling densities, herbaceous biomass, and duff depths. Large macroplots (~400 m^2) can be used to sample both logs to compute CWD loading and trees to compute canopy fuels. Sampling times can be shortened by employing an easier technique for estimating loading; the photoload method, for example, can be used to estimate loadings rather than actually measuring particle dimensions or destructively collecting fuels for weighing. Or, PI techniques can be used to estimate FWD on transects within a macroplot. And last, it may be more practical to sample fuels using FAP methods because many fuel components can be linked together in the same sampling unit.

The main limitation of the FAP sampling method is that there has yet to be a set of standardized operational FAP protocols for surface fuel sampling. Many fuel professionals are unfamiliar with the FAP technique and do not have the knowledge and expertise to create their own FAP methods. Moreover, FAP estimates tend to be unbiased but imprecise; variabilities of the loading estimates are often quite high. While sampled high variability often reflects realistic field conditions, it makes the detection of change across time or difference between areas using statistical techniques difficult, especially when compared against PI methods. Another limitation is that many fuel particles may extend to outside microplot boundaries making it difficult to cut particles without disrupting other fuel particles. And, unlike PI, it may take a high number of FAP microplots to obtain a reliable measure of loading (Keane and Gray 2013).

8.3.2.3 Distance Sampling

Another new method is *perpendicular distance sampling* (PDS) which samples logs using probability proportional to volume concepts (Ducey et al. 2013; Gove et al. 2012; Williams and Gove 2003). With PDS, the total volume of the logs on a landscape can be estimated from counts of logs at various sample points. Loading can then be estimated by multiplying volume by particle density (kg m^{-3}) estimates. PDS is named because a log is selected to the sample if a line from a sample point intersects the central axis of the log at a right angle and the length of this line is less than some limiting distance that changes along the log length in a manner that is based on the sampling design. There are many variants of PDS including the distance-limited protocol for PDS, which uses a fixed distance from the perpendicular line to estimate volume then loading (Ducey et al. 2013). Transect relascope, point relascope, and prism sweep sampling use angle gauge theory to expand on the PDS and line-transect method for sampling CWD (Bebber and Thomas 2003; Gove et al. 2005; Stahl 1998).

This method is most effective for measuring only one fuel component, CWD (Gove et al. 2012), but Ducey et al. (2008) demonstrated how PDS can be used to estimate other ecological attributes, perhaps finding a future use in FWD loading estimation. Several studies have compared traditional sampling techniques and PDS methods and variants to evaluate their performance, accuracy, and bias in measuring CWD (Bate et al. 2004; Delisle et al. 1988; Lutes 1999; Jordan et al. 2004; Woldendorp et al. 2004). Gove et al. (2013) compared PDS variants using simulation modeling and found unbiased estimators of CWD using all variants and the differences in variances were small across all variants so selection of the most appropriate variant depends on field conditions. Affleck (2008) merged PDS with PI sampling to create line intercept distance sampling to improve fuel sampling and got similar performances to PDS. Ståhl et al. (2010) merged critical length methods with PDS which appears to have improved CWD sampling. However, few studies have yet examined the performance of various sampling techniques for measuring across multiple fuelbed components, such as combinations of FWD and CWD, live and dead shrubs, and herbs on the forest. Because of this, there are few operational protocols that use PDS methods or variants for fuel inventory and monitoring.

8.3.2.4 Cover and Volume Sampling

An alternative to the above direct methods that measure fuel particle dimensions is applying the abundant methods that directly measure *canopy cover* in vegetation sampling efforts to fuel sampling, as opposed to visually estimating canopy cover as presented in Sect. 8.3.1.2. Canopy cover is directly measured using a suite of methods, techniques, and protocols for ecological inventories and research efforts (Krebs 1999; Mueller-Dombois and Ellenberg 1974), and some of these may potentially be applied to measuring fuel loading. Point sampling, for example, involves using a vertically placed rod of a small diameter to determine the particle that it contacts, and the number of contacts per particle type (i.e., fuel component) is then used to estimate cover. If applied to fuel sampling, the number of contacts can be correlated with the destructive sampling estimates of biomass. Measures of the height of each contact can be augmented with number of contacts to associate both cover and average height with loading (Catchpole and Wheeler 1992). Axelson et al. (2009), for example, used point and height methods to estimate shrub biomass for the western Australian karri forest. Line intercept techniques, where the length of intercept of plant parts are used to estimate cover, can also be used to estimate fuel loading.

Many other cover sampling techniques, such as relative frequency, could be modified for certain sampling environments to quantify loading (Lutes et al. 2006).

The problem with the modification of cover methods to estimate loadings is that canopy cover, regardless of how it's measured, may be poorly correlated with fuel loadings (Catchpole and Wheeler 1992). The depth of the litter layer, for example, is more correlated with loading than litter ground cover. Woody particle diameters, especially log diameters, are more important for computing loading than their projected cover. Many of these cover methods provide repeatable estimates with low bias compared to visual techniques. However, the use of cover methods to assess all fuel component loadings would not be recommended.

The volume method involves sampling the dimensions of a fuel particle or component to compute volume then multiplying volume by particle density or bulk density to get loading. An advantage of the volume method is that it can be used at particle, component, and fuelbed scale. Fuel component volume can indirectly calculated as discussed in Sect. 8.3.1.2 where the proportion measured cover (percentage cover divided by 100) is multiplied by height (m), sampling area (m^2), and bulk density (kg m^{-3}). Hood and Wu (2006) used the cover-volume approach to calculate loadings of masticated fuelbeds. Fuel component or particle dimensions can also be measured to directly estimate volume. Litter loading, for example, can be estimated by (1) measuring litter depths within a 1 m^2 microplot, (2) computing an average depth (m), (3) multiplying by sample unit FAP area (1 m^2) to calculate volume, and (4) calculating loading by multiplying volume (m^3) by bulk density (kg m^{-3}) and dividing by area of microplot (m^2). However, volume can also be used to estimate the mass of a fuel particle by (1) measuring particle dimensions (length, width, and depth), (2) estimating a volume by multiplying length, width, and depth, and then (3) multiplying particle volume by particle density to get dry weight. Loading can then be calculated by summing all particle dry weights over sample unit (FAP) area. As mentioned, bulk and particle density estimates for many fuel components can be found in the literature (Brown 1981; Keane et al. 2012b) or it can be estimated by destructively sampling a small proportion of the plots.

8.3.2.5 Destructive Sampling

As mentioned, *destructive sampling* involves removing fuel by clipping and collecting, drying the material, and weighing the dry mass of material. An alternative is to (1) collect and weigh the wet fuel in the field; (2) subsample that fuel to dry and weigh to estimate a moisture content, and then (3) use the subsampled moisture content to adjust the wet field weight to dry weight. Destructive sampling can be scaled for any sampling design or objective. Fuel particles can be collected individually, as a group (shrub or tree), or on FAPs. Destructive sampling almost always involves subsampling a fuel component or fuelbed so statistical methods are often required to summarize subsampled estimates to describe the sampling area. Often, destructive sampling is used to create predictive biomass equations for a fuel component or entity, such as a tree or shrub. This predictive equation can then be applied to inventory data to compute loading. Most destructive sampling is done for research rather than operational management inventory and monitoring.

8.3.3 Integrated Surface Fuel Sampling

Sampling projects are rarely designed using only one sampling approach or technique. The diversity of surface fuel components coupled with the constraints of limited resources always result in project-level sampling designs that compromise statistical rigor to ensure success by integrating the above techniques and approaches. Conventional standardized surface fuel sampling protocols nearly always recommend using PI techniques for woody fuel loading and volume approaches for litter, duff, shrub, and herb (Lutes et al. 2006). The photoload approach has been augmented with PI, fixed-area log sampling, and volume estimates for duff and litter (Keane et al. 2012b). Catchpole and Wheeler (1992) mention a sampling technique called "double sampling" where destructive techniques are used on a subsample of FAPs to calibrate loading estimates from visual techniques. Keane et al. (2012b) used double sampling for another reason—to adjust visual estimates using statistical regression. This melding of approaches, techniques, and intensities may aid in successful sampling designs, but the resultant loading estimates have different error distributions, variability, and usefulness for each fuel component. This makes evaluating fire model performance difficult when the uncertainties of loading estimates are different across fuel components.

8.4 Canopy Fuel Sampling

The five canopy fuel characteristics (*canopy base height* (*CBH*), *canopy height* (*CH*), *canopy bulk density* (*CBD*), *canopy fuel load* (*CFL*), and *canopy cover* (*CC*) in Chap. 4) can be estimated using any of four approaches. The first approach involves the destructive sampling of the canopy in vertical layers within a FAP (*destructive canopy methods*). Here, all canopy biomass is cut, dried, and weighed within a canopy layer for the plot area. This usually involves climbing or cutting trees, clipping their branches within a given canopy layer, clipping and sorting branch material into fuel components (needles, wood by diameter size class), and then drying and weighing the fuels in a laboratory. Reinhardt et al. (2006b), for example, sampled five forested stands in the western USA using this technique to describe CBD and CFL vertical distributions. This is obviously a time-consuming and costly method and is only done for research studies.

The second approach involves using various instruments or sampling schemes to indirectly measure canopy fuel variables (*indirect canopy methods*). In this approach, various specialized equipment or protocols are used to measure stand characteristics, such as gap fraction (percent of vertically projected canopy cover not containing canopy biomass), LAI, and canopy cover, and these simple measurements are then correlated to canopy fuel variables. Keane et al. (2005), for example, measured gap fraction using five different instruments for the destructively sampled plots of Reinhardt et al. (2006b) and then developed statistical models that predicted CBD from gap fraction for each instrument. These predictive relationships were then used by Poulos et al. (2007) to estimate CBD for stands in Texas, USA. Another alternative to this approach is using simple stand-level measurements to estimate canopy fuel variables. Alexander and Cruz (2014) created tables that show values for the five canopy fuel variables for various stand basal area and tree density classes for four western US forest types. These indirect techniques are relatively quick and cheap, but the instruments may be expensive, especially a terrestrial scanning LiDAR. However, the statistical models that predict canopy fuel characteristics from indirect measurement are generally not robust and are most accurate for stand types that are similar to the ones destructively sampled (Keane et al. 2005).

The most commonly used canopy sampling approach is estimating the canopy fuel variables using stand inventory data and modeling canopy biomass using allometric relationships (*allometric methods*). This technique uses an inventory of trees in a stand to compute the five canopy fuels variables. The inventory is often represented by a "tree list," which is a list of tree cohorts in the stand on a per area basis. Six attributes are usually measured for each tree cohort in most stand inventory protocols: tree density (trees per unit area), species, condition (live vs. dead), diameter breast height (DBH), height, and height to live crown base. There can be any number of tree cohorts in the tree list. The tree list is then used to compute the amount of canopy material in vertical canopy layers of specified thicknesses. This is often done by first computing burnable canopy biomass from the empirically derived allometric biomass equations for each crown fuel component (Brown 1978). This burnable canopy biomass is then distributed across the vertical crown length for each tree by assuming a crown shape and then using tree height and live crown base height to allocate biomass into each layer based on geometric analysis. The biomass is then summed across all trees for each layer and this sum is then divided by the volume of that layer (plot area multiplied by layer thickness) to calculate *CBD*. *CBH* and *CH* are calculated as the layer height at which the CBD exceeds or goes below a threshold value (Chap. 4). *CFL* is simply the sum of all burnable biomass over all layers divided by plot area. This technique is programmed into a computer application called FuelCalc (Reinhardt et al. 2006a).

One great advantage of the allometric method is that it can be used with any of the diverse stand inventories commonly conducted by natural resource management agencies. Moreover, the sampling techniques and methods for measuring trees using timber inventory techniques are widely known and many field crews are familiar with the protocols so training may be minimal. There are also many databases that contain tree lists that can be used to quantify canopy fuels characteristics; the US Forest Service's Forest Inventory and Analysis program has tree lists for thousands of plots across the USA. However, while this technique has been used for many fuel projects (Keane et al. 2006; Reeves et al. 2006) and is easily the most popular for sampling canopy fuels, there are some limitations. First, there are precious few studies that developed biomass equations for each of the crown fuel components (burnable canopy fuels; <3 mm), and most canopy biomass equations are for mostly western US forest species. Second, many of the assumptions in the method, such as crown shape and crown fuel distribution, may not be appropriate for some ecosystems and stand conditions. Third, some tree lists were created using stand inventory techniques that may be at an inappropriate scale. Plot-less sampling, for example, uses a prism or limiting distance sampling to determine which trees to sample, and these trees are usually above a certain breakpoint diameter resulting in few of the understory trees being sampled. The resultant tree list typically underrepresents the understory canopy biomass important to crown fire transition (Chap. 2). This means that overstory conditions will be summarized independently of understory conditions at stand level which ignores the importance of spatial autocorrelation in canopy fuel characteristics at smaller scales (Keane et al. 2012a). Sampling trees inside a FAP to create a tree list, and then using this plot-level tree list to compute canopy fuels provides for a better representation of canopy fuels than computing canopy fuels from averaged stand conditions.

The last canopy fuel sampling approach involves using a set of photos to visually estimate canopy fuel variables (*visual canopy methods*). Scott and Reinhardt (2005) developed a set of stereo photos of canopies from five western US sites in four different stand densities and calculated canopy fuel variables from destructive sampling at each site. These photos can then be compared to canopy conditions observed in the field to estimate the five canopy fuel variables along with other stand variables (basal area, tree density). Many of the newer photo series publications mentioned in Sect. 8.3.1.2 now have canopy fuel characteristics as attributes to the photos that are used to match with field conditions.

Estimating canopy fuel variables using field methods poses a dilemma to the fuels manager. The coarse resolution of crown fire modeling (Chap. 4) is often at odds with detailed sampling of canopy fuel characteristics. The coarser methods of indirect and visual canopy fuel sampling may provide sufficient resolution for the canopy fuel variables, and the more accurate and precise measurements gathered from the destructive and allometric methods may not match the coarse resolution of the canopy biomass in fire models. In fact, Reeves et al. (2006) created canopy fuels maps by quantifying canopy fuel variables from the allometric approach using the FuelCalc model but then had to adjust these precise measurements to use in the spatial fire prediction packages.

8.5 Challenges

The main challenge in fuel sampling is obtaining precise estimates of loading for each fuel component given the enormous spatial and structural variability across the different surface fuel components. With limited resources, it is simply impossible to sample to the same level of precision for all fuel components, and for some fuel components, it is incredibly difficult to obtain a precise estimate of loading without extensive sampling. The problem is that fuel component properties have unique spatial distributions that dictates the size and shape of the sampling unit. Small woody fuels (FWD) vary at scales that are much smaller than logs (CWD) (Keane et al. 2012a). Therefore, sampling designs must accommodate these spatial distributions along with the properties of the fuel component by using hierarchically nested sampling unit designs (e.g., nanoplots nested within microplots nested within macroplots).

Some fuel types are often ignored in most sampling projects for logistical, cost, and time reasons. Conifer seedling loading, for example, may comprise a significant portion of the fuelbed and contribute to fire ignition and spread (Fig. 3.4), yet few sampling designs include effective methods for sampling seedlings (Riccardi et al. 2007). Squirrel middens, animal scat, and pollen cones (Chap. 3) are other examples of fuel types that have few sampling methods and are rarely tied to fuel components (Ottmar et al. 2007).

Woody fuel loading should be stratified in statistically and ecologically appropriate size classes that still provide value in predicting fire behavior. Keane and Gray (2013) found the highest sampling uncertainty occurred when FWD were stratified by the conventional, nonuniform time-lag moisture sizes (e.g., 1, 10, 100, and 1000 h) rather than actually measuring particle diameters or measuring diameters to 1 cm size classes. As mentioned in Chap 3, the unbalanced Fosberg et al. (1970) size classes that get wider with larger particle diameters ignore subtle but important differences between species, degree of rot, and stand structure. Moreover, aggregating loadings of all log sizes into one class makes accurate decomposition predictions nearly impossible because of the great ecological importance of log size in various ecosystem processes (Harmon et al. 1986).

Four major biological factors are responsible for high levels of uncertainty in most sampling methods. First, wood density is highly variable both within and across the fuel particles, so the assumption of a constant density across all particles may be flawed. An assessment of density during sampling might improve loading estimates, but it would be difficult at this time to expect sampling crews to estimate particle density because there isn't any technology or standardized method as yet. Keane et al. (2012b) found high variability in wood density within a woody fuel component and an even higher variability within a sample site. Some woody fuel sampling protocols use the decay classes of Maser et al. (1979) to key to different wood densities (Lutes et al. 2009a; Lutes et al. 2006), but rarely are estimates of wood density actually measured in the field along with loading and rarely is the Maser et al. (1979) key applied to FWD. Second, woody fuel particles are not cylinders, but rather complicated volumes of highly variable cross-sections and contorted lengths (Chap. 3). Therefore, assumptions that woody fuel particle shapes can be approximated by frustums or cylinders using diameters and lengths may be oversimplified and techniques for measuring fuel diameters using rulers and gauges may be too coarse. Next, the high variability in fuel properties, specifically loading, within an area may often overwhelm targeted sampling precision. Often it is difficult to estimate fuel loading for coarser woody fuels to within reasonable targets without an impractical number of sample units. And last, fuelbeds are constantly changing over time; live fuels are constantly growing and depositing dead biomass on the decomposing necromass on ground (Chap. 6). The rates of biomass production, deposition, and decomposition change throughout the year because of plant phenology, climate, and disturbance. Sampling live fuels before the growing season, for example, may result in an underestimation of fuels that will burn during the fire season. Moreover, dead woody diameters are not static and change with weather conditions, often becoming thicker when wet, and cracked when dry, making diameter measurements difficult and further complicating the geometry used to estimate volume.

References

Affleck DLR (2008) A line intersect distance sampling strategy for downed wood inventory. Can J For Res 38(8):2262–2273. doi:10.1139/x08-056

Alexander ME, Cruz MG (2014) Tables for estimating canopy fuel characteristics from stand variables in four Interior West conifer forest types. For Sci 60(4):784–794

Arcos A, Alvarado E, Sandberg DV (1998) Volume estimation of large woody debris with a stereoscopic vision technique. In: Proceedings of the 13th Conference on Fire and Forest Meteorology, Lorne, Australia. International Association of Wildland Fire, pp 439–447

Axelson JN, Alfaro RI, Hawkes BC (2009) Influence of fire and mountain pine beetle on the dynamics of lodgepole pine stands in British Columbia, Canada. For Ecol Manage 257(9):1874–1882. doi:http://dx.doi.org/10.1016/j.foreco.2009.01.047

Baker WL (2009) Fire ecology in rocky mountain landscapes. Island Press, Washington, p 182

Bate LJ, Torgersen TR, Wisdom MJ, Garton EO (2004) Performance of sampling methods to estimate log characteristics for wildlife. For Ecol Manage 199:83–102

Bebber DP, Thomas SC (2003) Prism sweeps for coarse woody debris. Can J For Res 33:1737–1743

Brown JK (1970) A method for inventorying downed woody fuel. USDA Forest Service, Intermountain Research Station, General Technical Report INT-16. Ogden, UT, 16 pp

Brown JK (1971) A planar intersect method for sampling fuel volume and surface area. For Sci 17(1):96–102

Brown JK (1974) Handbook for inventorying downed woody material. USDA Forest Service, Intermountain Forest and Range Experiment Station, General Technical Report GTR-INT-16 Ogden, UT, USA, 34 pp

Brown JK (1978) Weight and density of crowns of Rocky Mountain conifers. United States Department of Agriculture, Forest Service Intermountain Forest and Range Experiment Station, Research Paper INT-197 Ogden, UT, USA, 56 pp

Brown JK (1981) Bulk densities of nonuniform surface fuels and their application to fire modeling. For Sci 27:667–683

Brown JK, Bevins CD (1986) Surface fuel loadings and predicted fire behavior for vegetation types in the northern Rocky Mountains. U.S. Department of Agriculture, Forest Service, Intermountain Forest and Range Experiment Station, Research Paper INT-358 Ogden, UT, USA, 9 pp

Brown JK, Oberheu RD, Johnston CM (1982) Handbook for inventoring surface fuels and biomass in the Interior West. USDA Forest Service, Intermountain Forest and Range Experiment Station, General Technical Report INT-129 Ogden, UT, USA, 48 pp

Busing R, Rimar K, Stolte KW, Stohlgren TJ, Waddell K (2000) Forest health monitoring. Vegetation pilot field methods guide. Vegetation diversity and structure down woody debris fuel loading. USDA Forest Service, National Forest Health Monitoring Program, Washington, D.C., USA, 54 pp

Catchpole WR, Wheeler CJ (1992) Estimating plant biomass: a review of techniques. Aus J Ecol 17(2):121–131. doi:10.1111/j.1442-9993.1992.tb00790.x

Delisle GP, Woodard PM, Titus SJ, Johnson AF (1988) Sample size and variability of fuel weight estimates in natural stands of lodgepole pine. Can J For Res 18:649–652

Dibble AC, Rees CA (2005) Does the lack of reference ecosystems limit our science? A case study in nonnnative invasive plants as forest fuels. J For 103(7):329–338

Ducey MJ, Williams MS, Gove JH, Valentine HT (2008) Simultaneous unbiased estimates of multiple downed wood attributes in perpendicular distance sampling. Can J For Res 38(7):2044–2051. doi:10.1139/x08-019

Ducey MJ, Williams MS, Gove JH, Roberge S, Kenning RS (2013) Distance-limited perpendicular distance sampling for coarse woody debris: theory and field results. Forestry 86(1):119–128. doi:10.1093/forestry/cps059

Fahnestock GR (1971) Weight of brushy forest fire fuels from photographs. For Sci 17(1):119–124

Fischer WC (1981) Photo guide for appraising downed woody fuels in Montana forests: Interior ponderosa pine, ponderosa pine-larch-Douglas-fir, larch-Douglas-fir, and Interior Douglas-fir cover types. USDA Forest Service, Intermountain Research Station, General Technical Report INT-197 Ogden, UT, USA, 133 pp

Fosberg MA, Lancaster JW, Schroeder MJ (1970) Fuel moisture response-drying relationships under standard and field conditions. For Sci 16:121–128

Gould JS, McCaw W, Cheney N, Ellis P, Matthews S (2008) field guide: fire in dry eucalypt forest: fuel assessment and fire behaviour prediction in dry eucalypt forest. CSIRO PUBLISHING, Canberra, ACT, Australia, 33 pp. ISBN: 0643102760

Gould JS, Lachlan McCaw W, Phillip Cheney N (2011) Quantifying fine fuel dynamics and structure in dry eucalypt forest (Eucalyptus marginata) in Western Australia for fire management. For Ecol Manage 262(3):531–546. doi:http://dx.doi.org/10.1016/j.foreco.2011.04.022

Gove JH, Williams MS, Stahl G, Ducey MJ (2005) Critical point relascope sampling for unbiased volume estimation of downed coarse woody debris. Forestry 78(4):417–431

Gove JH, Ducey MJ, Valentine HT, Williams MS (2012) A distance limited method for sampling downed coarse woody debris. For Ecol Manage 282(0):53–62. doi:http://dx.doi.org/10.1016/j.foreco.2012.06.014

Gove JH, Ducey MJ, Valentine HT, Williams MS (2013) A comprehensive comparison of perpendicular distance sampling methods for sampling downed coarse woody debris. Forestry 86(1):129–143. doi:10.1093/forestry/cps039

Hansen MH (1985) Line intersect sampling of wooded strips. For Sci 31(2):282–288

Harmon ME, Franklin JF, Swanson FJ, Sollins P, Gregory SV, Lattin JD, Anderson NH, Cline SP, Aumen NG, Sedell JR, Lienkaemper GW, Cromack K, Cummins KW (1986) Ecology of coarse woody debris in temperate ecosystems. Adv Ecol Res 15:133–302

Hood S, Wu R (2006) Estimating fuel bed loadings in masticated areas. In: Andrews PL, Butler BW (eds) Fuels management—how to measure success, Portland, Oregon, USA. U.S. Department of Agriculture, Forest Service, Rocky Mountain Research Station, Proceedings RMRS-P-41, pp 333–345

Jin S (2004) Computer classification of four major components of surface fuel in Northeast China by image: the first step towards describing spatial heterogeneity of surface fuels by images. For Ecol Manage 203:395–406

Jordan GJ, Ducey MJ, Gove JH (2004) Comparing line-intersect, fixed-area, and point relascope sampling for dead and downed coarse woody material in a managed northern hardwood forest. Can J For Res 34:1766–1775

Keane RE, Reinhardt ED, Scott J, Gray K, Reardon J (2005) Estimating forest canopy bulk density using six indirect methods. Can J For Res 35:724–739

Keane RE, Frescino TL, Reeves MC, Long J (2006) Mapping wildland fuels across large regions for the LANDFIRE prototype project. In: Rollins MG, Frame C (eds) The LANDFIRE prototype project: nationally consistent and locally relevant geospatial data for wildland fire management, 2006. USDA Forest Service Rocky Mountain Research Station, General Technical Report RMRS-GTR-175 Fort Collins, CO, USA, pp 367–396

Keane RE, Dickinson LJ (2007a) Development and evaluation of the photoload sampling technique. USDA Forest Service Rocky Mountain Research Station, Research Paper RMRS-RP-61CD Fort Collins, CO USA, 44 pp

Keane RE, Dickinson LJ (2007b) The photoload sampling technique: estimating surface fuel loadings using downward looking photographs. USDA Forest Service Rocky Mountain Research Station, General Technical Report RMRS-GTR-190 Fort Collins, CO, USA. 44 pp

Keane RE, Gray K, Bacciu V, Leirfallom S (2012a) Spatial scaling of wildland fuels for six forest and rangeland ecosystems of the northern Rocky Mountains, USA. Landsc Ecol 27(8):1213–1234. doi:10.1007/s10980-012-9773-9

Keane RE, Gray K, Bacciu V (2012b) Spatial variability of wildland fuel characteristics in northern Rocky Mountain ecosystems. USDA Forest Service Rocky Mountain Research Station, Research Paper RMRS-RP-98 Fort Collins, Colorado, USA, 58 pp

Keane RE, Gray K (2013) Comparing three sampling techniques for estimating fine woody down dead biomass. Intl J Wildland Fire 22(8):1093–1107. doi:http://dx.doi.org/10.1071/WF13038

Keane RE, Herynk JM, Toney C, Urbanski SP, Lutes DC, Ottmar RD (2013) Evaluating the performance and mapping of three fuel classification systems using forest inventory and analysis surface fuel measurements. For Ecol Manage 305:248–263. doi:http://dx.doi.org/10.1016/j.foreco.2013.06.001

Koski WH, Fischer WC (1979) Photo series for appraising thinning slash in north Idaho. USDA Forest Service Intermountain Forest and Range Experiment Station, General Technical Report INT-46 Ogden, UT, USA, 22 pp

Krebs CJ (1999) Ecological methodology, 2nd edn. Benjamin Cummings, New York

Loudermilk EL, Hiers JK, O'Brien JJ, Mitchell RJ, Singhania A, Fernandez JC, Cropper WP, Slatton KC (2009) Ground-based LIDAR: a novel approach to quantify fine-scale fuelbed characteristics. Int J Wildland Fire 18(6):676–685. doi:http://dx.doi.org/10.1071/WF07138

Lutes DC (1999) A comparison of methods for the quantification of coarse woody debris and identification of its spatial scale: a study from the Tenderfoot Experimental Forest, Montana. Master of Science Thesis, The University of Montana, Missoula, MT, USA, 55 pp

Lutes DC, Keane RE, Caratti JF, Key CH, Benson NC, Sutherland S, Gangi LJ (2006) FIREMON: fire effects monitoring and inventory system. USDA Forest Service Rocky Mountain Research Station, General Technical Report RMRS-GTR-164-CD Fort Collins 222 pp

Lutes DC, Benson NC, Keifer M, Caratti JF, Streetman SA (2009a) FFI: a software tool for ecological monitoring*. Int J Wildland Fire 18(3):310–314. doi:http://dx.doi.org/10.1071/WF08083

Lutes DC, Keane RE, Caratti JF (2009b) A surface fuels classification for estimating fire effects. Int J Wildland Fire 18:802–814

Maser C, Anderson RG, Cromack K, Williams JT, Martin RE (1979) Dead and down woody material. In: Wildlife habitats in managed forests: the Blue Mountains of Oregon and Washington. Agric Handb 553. US Department of Agriculture, Washington, DC, USA, pp 78–95

Maxwell WG (1976) Photo series for quantifying forest residues in the coastal Douglas-fir–hemlock type, coastal Douglas-fir–hardwood type. US Forest Service Pacific Northwest Forest and Range Experiment Station, General Technical Report PNW-51 Portland, OR, USA, 55 pp

McKenzie D, Raymond CL, Kellogg L-KB, Norheim RA, Andreu A, Bayard AC, Kopper KE, Elman E (2007) Mapping fuels at multiple scales: landscape application of the fuel characteristic classification system. Can J For Res 37:2421–2437

Morfin-Ríos JE, Alvarado-Deliestino E, Jardel-Peláez EJ, Vihnanek RE, Wright DK, Michel-Fuentes JM, Wright CS, Ottmar RD, Sandberg DV, Nájera-Díaz A (2007) Photo series

for quantifying forest fuels in Mexico: montane subtropical forests of the Sierra Madre del Sur; and temperate forests and montane shrubland of the northern Sierra Madre Oriental
Mueller-Dombois D, Ellenberg H (1974) Aims and methods of vegetation ecology. Wiley, New York, 234 pp
Nalder IA, Wein RW, Alexander ME, de Groot WJ (1999) Physical properties of dead and downed round-wood fuels in the boreal forests of western and northern Canada. Int J Wildland Fire 9(2):85–99
Nemec-Linnell AF, Davis G (2002) Efficiency of six line intersect sampling designs for estimating volume and density of coarse woody debris. BCMOF Vancouver Forest Region, Research Section, Technical Report No. 15 Nanaimo, BC, Canada 12 pp
Ottmar RD, Vihnanek RE (2000) Stereo photo series for quantifying natural fuels. Longleaf pine, pocosin, and marshgrass types in the southeast United States, vol VI, National Wildfire Coordinating Group National Interagency Fire Center, Boise
Ottmar RD, Sandberg DV, Riccardi CL, Prichard SJ (2007) An overview of the fuel characteristic classification system – quantifying, classifying, and creating fuelbeds for resource planning. Can J For Res 37:2383–2393
Poulos HM, Camp AE, Gatewood RG, Loomis L (2007) A hierarchical approach for scaling forest inventory and fuels data from local to landscape scales in the Davis Mountains, Texas, USA. For Ecol Manage 244(1–3):1–15
Reeves MC, Kost JR, Ryan KC (2006) Fuels products of the LANDFIRE project. In: Andrews PL, Butler BW (eds) Fuels management – how to measure success, Portland OR, 2006. USDA Forest Service Rocky Mountain Research Station, Proceedings RMRS-P-41 Fort Collins, CO USA pp 239–249
Reeves MC, Ryan KC, Rollins MC, Thompson TG (2009) Spatial fuel data products of the LANDFIRE project. Int J Wildland Fire 18:250–267
Reinhardt E, Keane RE, Brown JK (1997) First order fire effects model: FOFEM 4.0 User's Guide. USDA Forest Service, Intermountain Research Station General Technical Report INT-GTR-344. 65 pp
Reinhardt E, Lutes D, Scott J (2006a) FuelCalc: a method for estimating fuel characteristics. In: Andrews PL, Butler BW (eds) Fuels management—how to measure success, Portland, OR, pp 273–287. I U.S. Department of Agriculture, Forest Service, Rocky Mountain Research Station, Proceedings RMRS-P-41 Fort Collins, CO
Reinhardt E, Scott J, Gray K, Keane R (2006b) Estimating canopy fuel characteristics in five conifer stands in the western United States using tree and stand measurements. Can J For Res-Revue Canadienne De Recherche Forestiere 36(11):2803–2814. doi:Doi 10.1139/X06–157
Riccardi CL, Ottmar RD, Sandberg DV, Andreu A, Elman E, Kopper K, Long J (2007) The fuelbed: a key element of the Fuel Characteristic Classification System. Can J For Res 37:2394–2412
Sampson RN, Sampson RW (2005) Application of hazard and risk analysis at the project level to assess ecologic impact. For Ecol Manage 211:109–116
Sandberg DV, Ottmar RD, Cushon GH (2001) Characterizing fuels in the 21st century. Int J Wildland Fire 10:381–387
Scott JH, Reinhardt ED (2005) Stereo photo guide for estimating canopy fuel characteristics in conifer stands. USDA Forest Service Rocky Mountain Research Station, General Technical Report RMRS-GTR-145 Fort Collins, CO, USA, 47 pp
Sikkink PG, Keane RE (2008) A comparison of five sampling techniques to estimate surface fuel loading in montane forests. Int J Wildland Fire 17(3):363–379. doi:10.1071/Wf07003
Sneeuwjagt RJ (1973) Measuring forest fuels. Forests Department of Western Australia, Research Paper Number 9, Perth, Australia, 6 pp
Stahl G (1998) Transect relascope sampling – a method for the quantification of coarse woody debris. For Sci 44:58–63
Ståhl G, Gove J, Williams M, Ducey M (2010) Critical length sampling: a method to estimate the volume of downed coarse woody debris. Eur J For Res 129(6):993–1000. doi:10.1007/s10342-010-0382-3
Van Wagner CE (1968) The line intersect method in forest fuel sampling. For Sci 14(1):20–26

Vihnanek RE, Balog CS, Wright CS, Ottmar RD, Kelly JW (2009) Stereo photo series for quantifying natural fuels. Post-hurricane fuels in forests of the Southeast United States, vol XII, U.S. Department of Agriculture, Forest Service, Pacific Northwest Research Station, Gen. Tech. Rep. PNW-GTR-803. Portland, OR, USA, 53 pp

Vries PGD (1974) Multi-stage line intersect sampling. For Sci 20(2):129–134

Waddell KL (2001) Sampling coarse woody debris for multiple attributes in extensive resource inventories. Ecol Indic 1:139–153

Warren WG, Olsen PF (1964) A line intersect technique for assessing logging waste. For Sci 10(3):267–276

Williams MS, Gove JH (2003) Perpendicular distance sampling: an alternative method for sampling downed coarse woody debris. Can J For Res 33:1564–1579

Woldendorp G, Keenan RJ, Barry S, Spencer DR (2004) Analysis of sampling methods for coarse woody debris. For Ecol Manage 198:133–148

Chapter 9
Fuel Mapping

Knowing where things are, and why, is essential to rational decision making
Jack Dangermond, ESRI

9.1 Introduction

Since 1990, major advances in computer software and hardware have enabled development of spatially explicit fire growth models, thereby revolutionizing fire management decision support systems (Xiao-rui et al. 2005; Ball and Guertin 1992; Keane et al. 1998b). However, these complex spatial models demand detailed, high-resolution digital maps of surface and crown fuel characteristics to generate accurate and consistent fire behavior predictions (Pala et al. 1990). The commonly used FARSITE fire growth model, for example, requires five fuel layers to simulate surface and crown fire growth and intensity (Finney 1998). Early efforts at mapping fuels did not describe the physical aspects of the fuelbed, but rather interpreted resultant fire behavior if the fuels burned and how difficult it would be to suppress that fire, then mapped those attributes (Hornby 1936). With advancing computer technology, most fuel maps were developed to meet the input requirements of fire models (Keane et al. 1998a).

Fuel maps are now used in nearly all phases of fire management from planning to operational analysis at multiple organizational and spatial scales (Rollins 2009). Coarse scale fuel maps are integral to global, national, and regional fire danger assessment to more effectively plan, allocate, and mobilize suppression resources at weekly, monthly, and yearly evaluation intervals (Burgan et al. 1998; De Vasconcelos et al. 1998). Regional fuel maps are also useful as inputs for simulating regional carbon dynamics, smoke scenarios, and biogeochemical cycles (Kasischke et al. 1998; Leenhouts 1998; McKenzie et al. 2007), while finer scale subregional fuel layers are critically needed to rate ecosystem health (Keane et al. 2007), identifying fuel treatment locations (Agee and Skinner 2005), evaluating fire hazard and risk for land management planning (Hessburg et al. 2010), and aiding in environmental assessments and fire danger programs (Chuvieco and Salas 1996). However,

R. E. Keane, *Wildland Fuel Fundamentals and Applications*,
DOI 10.1007/978-3-319-09015-3_9

most fuel maps are used at finer scales, primarily for landscape assessments, because this is the scale at which most fires can effectively be simulated and managed (Heyerdahl et al. 2001). Landscape fuel maps are used to predict future spread of wildfires (Finney 2005), describe fire hazard and risk (Finney 2006), and portray fire severity (Karau and Keane 2010).

Creating wildland fuels maps is quite difficult, especially at landscape to regional scales, for a number of reasons (Arroyo et al. 2008; Keane et al. 2001). The lack of critical resources, such as limited geo-referenced fuel data and inadequate fuel classifications, coupled with a variety of ecological concerns, such as fuelbeds being hidden by the canopy and scale mismatches in field data, imagery, and analysis techniques, often complicate fuel-mapping efforts. Accurate fuels layers are costly to build because they require abundant field data, extensive expertise in a wide variety of spatial fields (remote sensing, geographic information system (GIS), fire and fuel modeling, image processing, vegetation mapping), and of course, a comprehensive knowledge of fuels (Keane et al. 2001). But most importantly, fuels are notoriously difficult to map because of their high variability and disparate spatial distributions across components (Chap. 6). This chapter first summarizes some critical mapping resources needed for nearly all mapping projects and then presents some general approaches used to map fuels for fire management at multiple scales. The challenges of fuel mapping are presented last to explain why most of today's fuel maps have some major limitations.

9.2 Fuel-Mapping Resources

9.2.1 Field Data

Field data are the most critical resource for mapping fuels, and collecting enough appropriate field data is often the most costly and time-consuming part of any mapping effort (see Chap. 8). Ground-based fuel sampling is literally the only way to realistically, accurately, and consistently describe the fuel characteristics being mapped (Keane et al. 2013) and it would be imprudent to attempt to map fuels without extensive field sampling. Geo-referenced field data are important for many reasons. First, field data provide important references for the mapped fuels classes because the data provide the only detailed descriptions of fuels (loading, classification category). Field plot data can also be used to describe polygons that can then be used as training areas in supervised classifications, or they can be used to describe unique clusters in unsupervised classifications (Verbyla 1995). More importantly, field data allow the development of statistical models for predicting fuel characteristics over space using ancillary biophysical spatial layers. Field data also provide a means for quantifying accuracy and precision of not only the fuel map but also the classification whose categories are being used as mapping units (Keane et al. 2013; Burgan and Hardy 1994). Plot data can be used to design and improve keys for the vegetation and fuels classifications being mapped. And most importantly, field data provide a means for interpreting fuel maps; inaccuracies or inconsistencies in mapping results can be explored using detailed plot data. A mapped shrub–herb category, for example, might be poorly mapped because the sampled cover of bare soil and rock was high on field plots.

9.2.2 Ancillary Spatial Data Layers

Fuel maps can be dramatically improved if supplementary spatial data are integrated into the mapping process (Keane et al. 2001). These ancillary spatial data often describe the biophysical environment to provide ecological context to the mapping process and to represent those processes that control fuel dynamics to increase predictive potential (Chap. 6). The most important ancillary GIS layer is the digital elevation model (DEM) that is used to describe the topography (e.g., slope, aspect, position) and indirectly represent the biophysical environment (e.g., climate). Many important topographic products can be derived from the DEM, such as slope position, stream corridors, and drainage basins (Skidmore 1989), to use as independent variables in statistical predictive models that create fuel maps. Moreover, it is possible to use the DEM as input to simulation models to create other biophysical layers, such as radiation, exposure, and microsite temperatures, and these new biophysical layers can be used to developed predictive relationships for mapping fuels (see Sect. 9.3.4). The DEM also is useful in delineating broad biophysical settings that can be used to stratify statistical modeling and fuel-mapping processes.

Perhaps the next most used ancillary data layers are digital maps of potential and existing vegetation classification systems, such as cover type, potential vegetation type, and structural stage maps (Menakis et al. 2000). Even though fuel loadings are rarely correlated to vegetation (Chap. 6), these maps are be important because they provide valuable context for assigning fuels to known settings, information on biophysical environment, and important linkages to other land management concerns. Vegetation layers are most useful if they were created across multiple scales using standardized, hierarchical classifications so that categories can be merged or split based on the ability of remote sensing to discriminate differences (Loveland et al. 1993; McKenzie et al. 2007). The most commonly used vegetation maps are ones that describe species composition (cover type), structure (vertical canopy layers), and some expression of potential vegetation (i.e., biophysical site; Menakis et al. 2000) because these three maps can be used to simulate vegetation development and therefore possibly fuel succession (Keane et al. 2006b).

Many other existing data layers have been used to map fuels. Spatial chronosequences of ecosystem characteristics, such as leaf area index (LAI), created from updated satellite imagery (e.g., MODIS), can be integrated in map development to quantify available biomass, represent fuel models, and correlate to many other fuel attributes (Rollins et al. 2004). Climate layers that integrate long-term weather into quantitative summaries that relate to fuel dynamics are also valuable ancillary layers

(Keane et al. 2001). Spatial soils data can also be used to describe the biophysical environment that can then be statistically related to fuel loadings or used in simulation models to create ancillary biophysical layers. Digital maps that describe social context (population density), transportation routes (roads, trails), utilities (power lines, gas lines), political (land ownership, management units), and ecological (stand maps, values at risk) resources can be used as references to characterize local to regional fuel differences and to stratify fuel assignments (Krasnow 2007).

The last set of ancillary data layers are those that are created from simulation modeling (Rollins et al. 2004; Keane et al. 2006a). Simulation modeling provides a platform to integrate disparate ancillary biophysical variables, such as climate, topography, and soils, into one comprehensive, integrated variable that may be more related to fuel attributes than the other variables separately. A potential evapotranspiration (PET) layer, computed from soils and climate data layers using an ecosystem model, may have a better relationship to fuel loading than the soils or climate data alone or together (Rollins et al. 2004). This simulation approach is discussed extensively in Sect. 9.3.4.

9.2.3 Fuel Classifications

A comprehensive fuel classification system is indispensable in fuel mapping because the classification's categories can serve as mapping units in the fuel map (Chap. 7). It is difficult to map loading, or any other fuel property, for each of the fuel components because of the high number of components and the fact that most components are difficult to map remotely, such as duff and litter, because they are hidden by higher fuel strata such as the forest canopy (see Sect. 9.4). Fuel classifications simplify the mapping process by providing a means to map all fuel components at once. Since most classifications were developed for specific fire applications, creating a map using a classification ensures that it will be useful in fire management. Finally, most fire managers are somewhat familiar with most existing fuel classifications, so mapping existing classifications eliminates the need for additional training to learn newly developed map units.

An ideal fuel classification for mapping should quantify a myriad of fuel characteristics (e.g., loading, size, bulk densities) for all fuel components at the appropriate mapping scale and resolution (Chap. 7). Fuel classification categories should be easily, accurately, and consistently identified in the field with comprehensive keys, and the classification should be related to other standardized vegetation and biophysical classifications (Keane 2013). The fuel classification should uniquely identify fuel types based on fuelbed characteristics, not on vegetation attributes or environmental descriptions, because the mapped categories must be easily validated in the field or using existing fuel data (Keane et al. 2013). Moreover, the classification structure should allow hierarchical aggregation and division so fuel categories can be tailored to match the strengths of the mapping approach, attributes of the remotely sensed products, resolution of available field data and imagery, and scale of eventual fire application. A link to other historical and current land-use maps is also desirable. Another desirable trait of useful fuel-mapping classifications is that the categories in the classification are easily and effectively discriminated by the diverse approaches used to map fuels (see Sect. 9.3).

Nearly all the fuel classifications mentioned in Chap. 7 have been used in fuel-mapping efforts. Perhaps the most mapped classifications are the fire behavior fuel models (FBFMs) which are needed to simulate wildfire in the USA. Reeves et al. (2009) created fine-scale (30 m) FBFM maps for both the Scott and Burgan (2005) and Anderson (1982) classifications for the contiguous USA. Root et al. (1985) mapped FBFMs for North Cascades in the US Pacific Northwest, while Peterson et al. (2012) produced FBFM maps for Yosemite National Park and Falkowski et al. (2005) for northern Idaho. McKenzie et al. (2007) mapped FCCS fuelbeds at 1 km for a national US scale and at 30 m for the Wenatchee National Forest, Washington, USA. Hawkes et al. (1995) mapped the fuel types in the Canadian Fire Behavior Prediction system for landscapes in British Columbia, Canada. The National Fire Danger Rating System (NFDRS) fuel types were mapped at a coarse scale by Burgan et al. (1998) for the USA and by Chuvieco and Salas (1996) for Spain.

There is a fundamental problem with using FBFMs as mapping units. The identification of FBFMs in the field is entirely subjective because it is based on an individual's perception of fire behavior under assumed weather rather than on actual measurements of fuel loadings (Chap. 7). Many field technicians find it difficult to consistently identify FBFMs on the ground because it requires knowledge of the fuel characteristics important to fire behavior, expertise in forecasting fire behavior in the field, and familiarity with the FBFMs. Even more important is that it is impossible to uniquely identify a FBFM from extant or legacy field data because a visual inspection of the fuelbed is absolutely essential for evaluating potential fire behavior (Anderson 1982). The FuelCalc program (Reinhardt et al. 2006) contains a routine that attempts to assign a FBFM from fuel loading data, but the routine has never been evaluated for accuracy and consistency. As a result, it is impossible to assess map accuracy for any of the FBFM classifications; one would have to observe fire behavior at a burning pixel to properly evaluate FBFM map accuracy. Reeves et al. (2009) addressed this subjectivity by holding calibration workshops attended by fire behavior specialists to evaluate fuel maps and adjust values where needed (Keane and Reeves 2011). And since most FBFMs quantify only a fraction of all dead and live biomass pools, they are rarely useful for most other fire applications such as smoke estimation and carbon cycling simulation.

9.3 Fuel-Mapping Approaches

Today's fuel maps are created by a complex merging of technologies and integration of analysis techniques (Arroyo et al. 2008). In general, there are four general approaches used to map fuels at multiple scales: field assessment, association, remote sensing, and biophysical modeling (Table 9.1). Early attempts at mapping fuels often used only one or two of these approaches, but as computing resources

Table 9.1 Summary of the approaches used to map fuels for fire management

Approach	Description	Advantages	Disadvantages
Field assessment	Using ground-based surveys and field reconnaissance to assign fuel attributes across an area	Mapping actual observations; no introduced modeling or analysis error; limited number of steps; easily modified and refined; easily augmented with field sampling	Costly and time-consuming; inconsistency between mappers; somewhat subjective; bias toward mountainous terrain
Association	Assigning fuel attributes to categories in extant classifications, often vegetation classifications	Simple, direct, and easy; no need for additional mapping; most extant classifications are well known and easily understood; can assign many fuel attributes to one category; can use many classifications to fine-tune fuel assignments; create robust maps useful for other natural resource applications	Fuels are often unrelated to vegetation categories; scale and resolution of extant classification does not match scale of fuel data or subsequent application; high redundancy in fuel attributes across extant classification categories
Remote sensing	Correlating remotely sensed imagery with fuel characteristics	Readily available; provides snapshot of existing conditions; well accepted and long history of resource mapping using remote sensing products; many products available at different resolutions and detail	Fuels often are uncorrelated to imagery signals; scale of imagery may not match scale of fuel distribution; requires extensive expertise in remote sensing, GIS, statistical modeling, and wildland fuel science; difficult to understand reasons for mapping successes or failures
Biophysical modeling	Calculating or simulating biophysical gradients to correlate to fuel attributes	Relating those processes that control fuel dynamics to fuel mapping; provides context for interpreting fuel maps; can often simulate environmental gradients at multiple time and space scales; can be used to map many other ecological characteristics to augment fuel mapping	Best describes potential rather than existing fuel conditions; fuel attributes often are uncorrelated to biophysical variables; difficult to decide the type, detail, and scale of the biophysical gradient that best represents fuel dynamics; scale of simulated gradient may not match scale of fuel distribution or fuel processes; requires extensive expertise in ecosystem modeling, GIS, statistical modeling, and wildland fuel science; demands extensive data for initialization and parameterization

improved, mapping expertise increased, and extensive spatial ecological data sets became available, most of today's fuel-mapping efforts integrate these multiple technologies to get the best possible fuel maps (Keane et al. 2001). Therefore, these approaches should not be considered methods *per se*, but rather a set of general strategies to map fuels.

Several analysis methods were not included as approaches in this chapter because they are used across most of the four mapping approaches. The most important and most commonly used analysis method is statistical modeling, where advanced statistical techniques, such as multiple regression analysis, generalized linear modeling, and regression trees, are used with field and spatial data to create empirical models that are then employed to build fuels maps (Miller et al. 2003). Another exciting branch of spatial analysis is the integration of expert knowledge into numerical analysis to develop fuel maps (Keane and Reeves 2011); the vast knowledge and expertise of fire professionals can be used to develop and test fuel maps using a wide variety of computing technology, such as expert systems, neural networks, and artificial intelligence (Krivtsov et al. 2009).

9.3.1 Field Assessment

Field assessments involve traversing a landscape on the ground and recording fuel conditions using data recorders, notebooks, or paper maps (Arroyo et al. 2008). Conditions in the field are assessed using a diversity of methods that include actual sampling of the fuel (Chap. 8), recording a category in a fuel classification category (Chap. 7), or describing the fuel type using vegetation, disturbance, and site characteristics. The observed conditions are then assigned to polygons on a photo or map. Few fuel maps were created using this approach, and of those that were, they were mostly for fine-scale, small-area projects. The exception was Hornby (1936), who remarkably mapped more than 6 million ha in the northern Rocky Mountains using more than 90 Civilian Conservation Corps (CCC) workers. These crews walked, rode, or drove through national forests in Montana and Idaho of the USA and described fuel conditions by coloring polygons on maps. But, instead of actually recording fuels loadings, the CCC crews mapped two categorical fire behavior descriptors that were inferred from the fuel conditions: resistance to control and rate of fire spread. The fuel classification used by Hornby (1936) was only useful for one fire management purpose, suppressing wildfires. Many employed the Hornby (1935) methods to other parts of the country (Abell 1937; Banks and Frayer 1966; New Jersey Department of Conservation and Development 1942) (Chap. 1).

The primary advantage of the field survey strategy is that fuels are mapped from actual conditions observed on the ground (Table 9.1). Mapping error is limited to erroneous fuel-type assessments or improper stand delineations on paper maps and no error is introduced from inappropriate statistical modeling or data analysis. Fuel assignments can be subjectively adjusted based on the observers' knowledge of the fuel complex, of how fire burns the fuel complex, and of how fire behavior models simulate burning in the fuel complex. Observers are easily able to visually assess highly variable fuel conditions across large areas to estimate an average or representative value providing there is extensive training. Any special conditions that arise in the field, such as the identification of a new fuel type or the elimination of a rare fuel type, can be easily integrated into the mapping scheme. And this approach can easily be augmented with field sampling to increase accuracies and map detail. It can also be scaled to specific projects creating anywhere from high-resolution maps for small areas to coarse-resolution maps for large regions.

The great amount of effort involved in a successful field approach would probably preclude its use in most large-scale operational fuel-mapping projects today. The majority of time and money spent on any fuel-mapping effort is usually in field assessments of fuel conditions so assessing the entire map area would be impractical. Another drawback is that there are always inconsistencies between field observers because of differences in their expertise and knowledge of fuels and fire (Sikkink and Keane 2008). And there is a sampling bias toward mountainous terrain since most of the reconnaissance mapping efforts are done from observation points on high, burned-over vistas, so locations not directly seen from these observation areas were probably mapped with less accuracy (Brown and Davis 1973). This approach would be more valuable if it were integrated with field sampling to create the field reference datasets to augment with other fuel-mapping approaches.

9.3.2 Association

In the association approach, fuel maps are developed by assigning fuel attributes to the categories or mapping units of maps of other land classifications, similar to the associative fuel classification (Chap. 7) and associative fuel-sampling (Chap. 8) approaches. There are a number of readily available, well-known spatial data layers of vegetation, topography, and land use that can be used either alone or in combination to associate fuel characteristics to each classification category or combination (McKenzie et al. 2007). In the association process, fuel attributes are usually quantified or selected from a synthesis of field data across extant classification categories. These fuel attributes are then assigned to that category to create the fuel map from the existing map. Satellite imagery and other remotely sensed products are better suited for differentiating between vegetation types than fuel types (Keane et al. 2001). Keane et al. (1998a), for example, overlaid maps of vegetation and topography classifications with plot-level geo-referenced FBFM assessments and, for each vegetation and topography class combination, they assigned the modal FBFM of all field plots within that combination. A fuel type group map was created by averaging fuel loadings for each of eight fuel components for all USFS Forest Inventory and Analysis plots in each forest-type group category (Keane et al. 2013). This approach may also be used with expert knowledge techniques that assign fuel-classification categories to other map categories using the experiences of fire professionals (Keane and Reeves 2011) or statistical analysis of field data to build empirical models that assign fuel characteristics to other classification categories

(Reeves et al. 2009). The associative approach is easily the most commonly used approach for developing fuel maps.

Examples of this approach can be presented by spatial scale. Coarse-scale imagery is often used to discriminate broad vegetation types or land cover classes, and these classes sometimes correlate with fuels because vegetation categories are so broad they generally have unique fuel characteristics. Burgan et al. (1998) used Omernik (1987) ecoregions and the Loveland et al. (1991) AVHRR land-cover classification to develop an NFDRS fuel model map of the conterminous USA. Landsat imagery was used to map vegetation on 100 million ha in Alaska, and then fuel models were assigned to each vegetation category (Willis 1985). McKenzie et al. (2007) mapped FCCS fuelbeds to vegetation and disturbance classification categories, and the FCCS fuelbeds of Ottmar et al. (1994) were assigned to combinations of vegetation cover and structure types for the Interior Columbia Basin Ecosystem Management Project (Quigley et al. 1996). Menakis et al. (2000) expounded on the "vegetation triplet" approach where fuel models or classes are assigned to categories in three classifications: potential vegetation, vegetation composition, and vegetation structure. Jain et al. (1996) intensively sampled fuels for all categories of a forest-type map created from Linear Image Self Scanning (LISS II) imagery to create a fuel map for Rajaji National Park in India. In Canada, the Canadian Forest Fire Behaviour Prediction System (FBP, Forestry Canada Fire Danger Group 1992) fuel types were assigned to vegetation categories on maps created from Landsat Multi-Spectral Scanner (MSS) data for Wood Buffalo National Park (Wilson et al. 1994), Quebec (Kourtz 1977), British Columbia (Hawkes et al. 1995), and Manitoba (Dixon et al. 1985).

The association approach is used for many reasons. The most common reason is that it is relatively easy, quick, and economical to create fuel maps from other maps because they can be done by anybody for any location where there is an associative map. There are many vegetation classification maps available to associate fuel characteristics (Anderson et al. 1998; Grossman et al. 1998), and most people can easily identify the vegetation-type categories of these classifications in the field. There are also many field data sets that contain assessments of these extant classification categories at the plot level that can augment fuel mapping. Since extant classification maps are used extensively in resource management, the assignment of fuel attributes are easily understood by managers, and the resultant fuel maps can be linked to other resource concerns. Many fuel attributes can be assigned to an extant category allowing the creation of many types of fuel maps, such as surface fuel maps and canopy fuel maps (Keane et al. 2000). Finally, associative maps often provide a context for interpreting fuel distributions across a landscape. For example, it is helpful to know that a polygon was assigned a needle and litter FBFM because it was a ponderosa pine stand.

The major disadvantage of association in fuel mapping is that fuels are not always correlated with vegetation characteristics or land-use categories so statistical relationships between fuel and the associated layers may be too weak to develop useful predictive models (Chap. 6). An example of this lack of relationship is the redundancy of fuel classes across the associated mapped classification classes. For example, there were as many as four different FBFMs found in the many of the combinations of vegetation structure, species composition, and topographic settings classes for maps of the Selway-Bitterroot Wilderness Area, USA. (Keane et al. 1998a). Stand disturbance history, biophysical setting, and vegetation structure are significant factors governing fuel characteristics so they should be incorporated into the fuel model assignment protocols. Also, the scales of the base classifications may not match the scale of the fuels being mapped or the sample design of the field data used in the mapping (Keane et al. 2006a). The vegetation categories in the Society of American Foresters (SAF) cover-type classification used in the FOFEM model, for example, are so broad for some cover types that they encompass a wide variety of fuelbed conditions that overwhelm important local differences (Schmidt et al. 2002). Other disadvantages are compounding errors occurring when the error inherent in the original base classifications is combined with errors in the fuels classifications and errors in fuel class assignment (Keane et al. 2013).

9.3.3 Remote Sensing

Remote sensing approaches attempt to correlate remotely sensed imagery with fuel characteristics using statistical modeling to create a fuel map (Keane et al. 2001; Lanorte et al. 2011). The imagery can be from any number of passive and active sensors. Passive sensors include digital aerial photography (Oswald et al. 1999), Landsat Thematic Mapper (TM; Brandis and Jacobson 2003), Advanced Spaceborne Thermal Emission and Reflection Radiometer (ASTER; Falkowski et al. 2005), and hyperspectral (Jia et al. 2006), while active sensors are usually LiDAR (Andersen et al. 2005) and radar (Bergen and Dobson 1999). These sensors can be mounted on any number of platforms including fixed wing aircraft, helicopters, and satellites to obtain a wide range of resolutions and detail (Xiao-rui et al. 2005). Passive sensors usually measure the reflectance of light in a narrow band of the electromagnetic spectrum, and some of these sensors, such as Landsat's TM with a 30 m pixel size, create multiple data layers that represent the reflectance from multiple spectral bands. Hyperspectral imagery, such as Airborne Visible InfraRed Imaging Spectrometer (AVIRIS), Hyperion, and HYDICE, may have more than 50 different spectral reflectance layers. Active sensors, such as LiDAR, consist of a cloud of point measurements of return times and signal strengths that are then used to statistically model height and loading (Riaño et al. 2003).

The central assumption of the remote sensing approach is that there is a correlation between fuel characteristics and the remotely sensed data signal. Fuel attributes, such as loading, canopy bulk density (*CBD*), or classification categories, either computed from legacy plot data or measured directly on geo-referenced plots, are related to the reflectance values of the plot location using simple to complex statistical modeling. Two general statistical methods are used to create fuel maps. In the supervised classification technique, statistical models that directly predict fuels information are built from the reflectance values of the imagery and the field data. Then, fuel maps are then created by employing the developed predictive relationships across all pixels. In the unsupervised classification technique, the reflectance values of all pixels are used in advanced statistical clustering methods to create unique spectral "signatures" and then various statistical techniques are used to assign the geo-referenced plot information to the mapped spectral signatures. Numerous other data layers and spatial information can be augmented with the spectral imagery to improve both the unsupervised and supervised statistical analyses (see Sect. 9.3.2).

Many types of fuel maps have been created using passive satellite imagery, mainly from Landsat satellite sensors. The majority of fuel-mapping efforts used from Landsat MSS and TM imagery to map surface fuel classification categories. Kourtz (1977) used Landsat MSS data to map fuel models in Canada. Salas and Chuvieco (1994) classified Landsat TM imagery directly to 11 of Anderson's (1982) fuel models, then assigned vegetation categories to each fuel model to compute fire risk on a large landscape in Spain. An Anderson (1982) FBFM map was classified directly from TM imagery of Camp Lejeune, North Carolina, for simulating prescribed fires with FARSITE (Campbell et al. 1995). However, the highest successes are when total living and dead biomass were directly mapped to spectral signatures. Direct biomass imagery mapping is more accurate for grasslands and shrublands (Chladil and Nunez 1995; Millington et al. 1994; Friedl et al. 1994), but less certain when assessing surface fuels in forested ecosystems because of the canopy obstruction problem (Elvidge 1988; see Sect. 9.4). Merrill et al. (1993) estimated living grassland biomass in Yellowstone National Park using regression models on bands 4, 6, and 7 from Landsat MSS) imagery. Using TM imagery, Peterson et al. (2012) directly mapped 1-, 10-, and 100-h loadings in Yosemite National Park USA and Brandis and Jacobson (2003) mapped total fuel loads in Australia. Large-scale aerial photography and aerial sketch mapping have been used successfully to estimate natural and slash fuel distributions in a variety of forested settings in Canada (Belfort 1988; Morris 1970; Muraro 1970).

Other imagery has been successfully used in fuel-mapping efforts. At fine scales, Lasaponara and Lanorte (2007a) used QuickBird high-resolution imagery (2.9 m) to map fuel types in Italy. ASTER imagery, having higher spectral (15 bands) and spatial (15 m) resolution than Landsat TM (7 spectral plus a panchromatic band, 30 m spatial resolution), was used to map Mediterranean fuel types in southern Italy (Lasaponara and Lanorte 2007b) and the 13 Anderson (1982) FBFMs in Idaho, USA (Falkowski et al. 2005). Root and Wagtendonk (1999) used hyperspectral imagery to map fuels in Yosemite National Park, USA, while Jia et al. (2006) used AVIRIS hyperspectral imagery to map canopy fuels. Active remote sensors such as Synthetic Aperture Radar (SAR) that propagate pulses of electromagnetic radiation and detect the reflective backscatter have shown promise for mapping stand biomass (Rignot et al. 1994) so they may be useful for estimating surface fuel models, crown bulk densities, and canopy dimensions. In Yellowstone National Park in the USA, Saatchi et al. (2007) mapped canopy fuel characteristics and Huang et al. (2009) mapped CWD using SAR and other ancillary data layers. Keramitsoglou et al. (2008) fused hyperspectral imagery with ASTER to map fuel types in Greece.

Airborne LiDAR appears to be the most promising remotely sensed product for the mapping of fuel properties, especially canopy fuel attributes, because it describes the vertical profile of the fuelbed. LiDAR estimates distance to an object by measuring the time delay between the transmission of a pulse of light and the detection of the reflected light from a target. This process, in a vegetative setting, can result in millions of points in an area that describe the fuel strata. The point distances can be used to calculate elevations to map a fuelbed in three dimensions if the spatial density of laser measurements is high. The distribution of elevations can be used as a signal to map fuels and the strength of the return signal is also useful for determining the surface condition that may be related to certain fuel types. Some have used LiDAR to map surface FBFMs with some success (Mutlu et al. 2008), but the real strength of LiDAR is in the mapping of canopy fuels (Andersen et al. 2005; Erdody and Moskal 2010) because the number of LiDAR distance measurements within the canopy profile is often correlated to CBD and canopy base height (CBH; Riaño et al. 2003). However, LiDAR also has its problems. While it can accurately produce a canopy height profile, it has limited ability in differentiating the material that reflected the laser intercept; it is difficult to tell if the piece of biomass hit by the laser was a leaf, twig, or log. The canopy obstruction problem is also a factor in that upper canopies obscure lower canopy strata and thereby collect a disproportionate number of LiDAR hits. Loadings for those fuel components that contain the majority of dead biomass, logs, litter, and duff are also difficult to sense from LiDAR because their size or depth is nearly impossible to measure using LiDAR.

There are advantages to using a remote sensing approach (Arroyo et al. 2008). First, unlike all other approaches, remotely sensed data provide a spatial description of existing landscape conditions and act as a snapshot of the landscape. As such, these data can be useful for the detection of changes in fuel conditions through time and space. Most imagery products are easy to obtain but their cost is highly variable ranging from free to quite expensive. Remotely sensed imagery can be obtained for a wide variety of resolutions allowing appropriate scaling of the imagery to fuel component distributions.

Logistical concerns, however, may limit many remotely sensed fuel-mapping projects. Expertise in image processing, GIS analysis, and statistical modeling is rare and expensive, and combined with expertise in fuel science and fire behavior modeling, the number of people qualified for fuel-mapping projects are scarce. Absolutely critical to remotely sensed fuel-mapping projects are surface and canopy fuel data which are often limiting in most areas. The analysis of the imagery also demands high computing resources which may be restrictive for many fire managers. Finally, many of the remotely sensed products, such as LiDAR, ASTER, and SAR, may be too expensive for operational fuel mapping across large domains and require specialized expertise in data processing.

There are also important ecological limitations of remote sensing approaches for fuel mapping. As mentioned, some fuel component attributes, such as CBH, FWD, and herbs, are obscured by the canopy in most forest and some shrubland ecosystems (Keane et al. 2001). Even if the fuel components were visible from above, the remotely sensed imagery probably would probably have low correlation to many attributes that are being mapped, such as loading, because of the mismatch in scales. Logs and FWD are too small to be sensed by most imagery products with 30-m pixel resolution, yet they comprise the majority of loading in some environments. Duff and litter loading, as another example, depends on their depth on the ground, and this depth is rarely correlated to imagery signals (Asner 1998). Most imaging sensors were designed to differentiate vegetation characteristics, so vegetation conditions may often overwhelm any fuel signal, and most fuel components, such as woody fuels, have similar reflective properties making it difficult for their differentiation.

Another limitation is that it is often difficult to quantify fuelbed characteristics for each component with only one unique spectral signature, unless, of course, a fuel classification is being mapped, but then few fuel classifications are highly correlated to imagery (Keane et al. 2013). Conversely, if fuel components are mapped separately, there is a good chance that each component map will be spatially incongruent or inconsistent, and impossible combinations may result. And, since fuel components are spatially distributed at different scales, using only one imagery product with one resolution ensures some fuels may always be mapped at an inappropriate scale (Keane et al. 2012a; Chap. 6); fine fuels important for fire spread are too small to be detected accurately by most passive imagery products. It is also difficult to detect the vertical distribution of fuels with passive imagery; the sensed FWD might actually be suspended above the ground.

9.3.4 Biophysical Modeling

This last approach relates fuel attributes to measured or simulated biophysical gradients using statistical modeling. Biophysical gradients describe those ecological phenomena that may directly or indirectly influence fuel dynamics (Chap. 6), such as climate, productivity, and disturbance. Spatial data representing these gradients can be (1) measured directly, such as climate, soils, and topography, (2) measured indirectly by correlating with imagery, or (3) simulated using biophysical models. The direct and indirect gradients are often used as inputs into biophysical models to create additional gradients.

Ecosystem models have vastly improved over the past two decades and there are a wide variety of models for application at coarse (e.g., MAPPS, Lenihan et al. 1998), regional (e.g., BIOME-BGC, Thornton et al. 2002), and fine scales (e.g., FireBGCv2, Keane et al. 2011). These models simulate those ecosystem processes known to govern fuel dynamics and these simulated processes can then be mapped and used to predict fuel characteristics across space. Relationships between biophysical processes and organic matter accumulation and decomposition, for example, can be used to predict fuel characteristics (Gosz 1992; Ohmann and Spies 1998). Rollins et al. (2004) developed a prototype system to link remote sensing, gradient modeling, and ecosystem simulation into a package for mapping those characteristics important to land management, and then used the system to map FBFMs (Keane et al. 2006a). Biophysical layers can be topographical (e.g., elevation, aspect, slope), biological (e.g., successional stages), geological (e.g., soils, landform), or biogeochemical (i.e., evapotranspiration, productivity, nutrient availability). Kessell (1976) used seven biophysical gradients based on topography and vegetation to spatially predict fuel models and loadings in Glacier National Park, Montana. Habeck (1976) sampled fuels and vegetation in the Selway-Bitterroot Wilderness Area of Idaho and related fuel loadings to stand age and moisture–temperature gradients. Keane et al. (1997) developed a protocol for mapping surface fuels from several biogeochemical and biophysical variables using an extensive network of field plots, and later used those techniques for mapping canopy fuels (Keane et al. 2006a).

The value of this approach is that simulated environmental gradients provide an ecological context in which to understand, explore, and finally, predict fuel dynamics. Low fuel loadings in a stand, for example, may be explained by low precipitation, high evapotranspiration, and low productivity. Furthermore, environmental gradients can quantify those important ecosystem processes that correlate with fuels, such as decomposition, to provide a temporal and spatial framework for creating dynamic fuels maps. Climate change effects on spatial fuel loadings can be easily computed by evaluating changes in environmental gradients under the new climate (Keane et al. 1996). Most environmental gradients are scale-independent, meaning the same gradients might be useful to predict fuel characteristics across many spatial scales, but the range, distributions, and strengths of the relationships might change. These models can also be used to update fuel maps by simulating deposition and decomposition processes to see how the fuels have changed over the life of the map. And once biophysical layers are developed, they may be used by land management agencies for many management applications (Keane et al. 2002).

One major problem with this approach is that biophysical gradients do not provide a comprehensive description of existing biotic conditions so remotely sensed data are often needed to spatially portray the current fuel conditions. Another disadvantage is that this approach requires abundant field data, complex ecosystem models, and intensive statistical analyses requiring extensive expertise in ecological sampling, simulation modeling, and statistical examination. Ecosystem models demand comprehensive initialization, parameterization, calibration, and validation to be useful, and this often requires extensive data, time, expertise, and computing resources. Biophysical settings are inherently difficult to map because they represent the complex integration of long-term climatic interactions with vegetation, soils, fauna, and disturbance (Barrett and Arno 1991; Habeck 1976; Keane et al. 1996b). Moreover, identification of those biophysical processes critical to fuel dynamics is difficult because most are unknown or unquantifiable, and they are difficult to identify in the field because of their temporal aspect. Many biophysical layers may have limited value for mapping fuels because of interacting factors and they are often correlated with other biophysical processes. And last, all biophysical gradients affect fuel processes at different scales so it is important that the biophysical layers are created at the most appropriate scales that influence fuel properties.

9.3.5 Integrating Approaches

Most mapping projects integrate all approaches to create state-of-the-art fuel maps. Peterson et al. (2012) statistically modeled live and dead woody fuel component loadings using regression classification procedures with a suite of climate, topography, imagery, and fire history-independent variables. Varga and Asner (2008) merged LiDAR with hyperspectral imagery to map surface fuels in Hawaii. A knowledge-based system of neural networks was used to search for unique fuel patterns on a large landscape in Portugal from land-use, vegetation, satellite imagery, and elevation information (Vasconcelos et al. 1998). Pierce et al. (2012) used intensive field sampling to describe surface fuels for spectral clusters in an unsupervised approach and correlated canopy fuel characteristics to topography (elevation, slope, aspect) and Landsat TM imagery using Random Forests statistical modeling. And, in the most extensive fuel-mapping effort in the USA, Reeves et al. (2009) mapped canopy fuel attributes (*CBD, CBH*) for the contiguous USA by creating regression models from Landsat TM reflectance imagery, biophysical gradients simulated by an ecosystem process model, and topographic variables calculated from the DEM. They also mapped four surface fuel classifications using an associative approach where categories were assigned to combinations of vegetation cover, structure, and biophysical classifications using statistical modeling and expert opinion. The merging of multiple approaches has resulted in some of the most useful and accurate fuel maps.

9.4 Challenges

The accuracy of fuels maps varies widely, but generally, most fuel maps have low accuracies. When accuracy assessments were reported, they usually ranged between 5 and 85 % correct, regardless of fuel-mapping approach or integrative strategy (Keane et al. 2013). Fuel map accuracies often reflect the approaches used to create the maps; maps created with the associative approach, for example, tend to have the same accuracies as the core maps used to associate fuel attributes. Low map accuracies, however, don't always mean the fuel map is worthless, especially considering the high variability and complexity of fuels. Alternative management strategies can be effectively compared by assessing the relative differences in fuel conditions between sites in fuel maps with precision. Low fuel map accuracies may be a result of a number of inherent sampling and analysis errors that are out of the mapper's control, such as (1) scale differences in field data and mapped elements, (2) improper geo-registration, (3) erroneous field identification or measurement of a mapped attribute, (4) improper use of vegetation or fuels classifications, (5) mistakes in field data entry, (6) differences in sampling error across fuel components, and (7) inappropriate fuel-sampling methods and designs. However, the main reason for low fuel map accuracies probably lies in the ecology of fuels rather than in the limitations of the approaches and data used to map them.

Several ecological reasons are to blame for the low accuracies in most fuel maps. As with other fuel applications, the high variability of fuel characteristics in space and time across the diversity of components compromises most fuel-mapping efforts (Chap. 6). In a validation of the LANDFIRE fuel maps, Keane et al. (2013) found that the inability of a fuel classifications' category to uniquely quantify fuel loadings was the biggest reason for poor mapping results. This inability to predict fuel loadings was mainly because of the high variability of loadings across components within a classification category (Chap. 6). High variability of loadings across classification categories is often because fuel components vary at different scales and are uncorrelated with each other (Keane et al. 2012b). Keane et al. (2000) hierarchically assessed accuracy of vegetation and fuel maps by quantifying error in the field data, vegetation and fuel classifications, and found more than 20 % of map error resulted from the inherent variability of fuel components attributes sampled at the stand-level. This high loading variability is also because fuel components are spatially distributed at different scales and accumulate at different rates (Chap. 6). In summary, the high variability of fuel attributes, especially loading, often overwhelms any spectral or biophysical signal used for mapping, resulting in inadequate discrimination of fuel classification categories and attributes.

Stand disturbance history, expressed as time since last fire for an example, is perhaps the single most important factor dictating fuel bed characteristics (Chap. 6) yet there are few ancillary spatial data sources that describe stand history that can be used in fuel mapping. Vogelmann et al. (2011) use fire severity maps to update the LANDFIRE vegetation and fuels data layers, but there are few comprehensive maps of other disturbances. Past fires both reduce fuel component loadings by consumption and increase loadings by causing plant mortality (Chap. 6). Insects, diseases, and wind often increase fuel loadings disproportionately across components. Without a spatial description of the timing, severity, and extent of past disturbances, it will always be difficult to map fuels.

There may be other logistical reasons for poor map accuracies. The biggest limitation in most fuels mapping is the lack of timely, dependable, geo-referenced field data describing existing fuels conditions. Few comprehensive standardized fuel-sampling efforts have created the databases needed for fuel-mapping efforts. For those projects where fuels were actually measured, inadequate training in fuel model assessment and fuel measurement techniques resulted in questionable field estimates (Keane et al. 1998b). Fuel characteristics (e.g., surface fuel model, crown fuels, stand height) should not be mapped independently or illogical combinations will inevitably result. All fuel layers must be developed and mapped in parallel so they are spatially congruent and consistent.

Low fuel map accuracies may be improved by employing newer methods and better technology, but there are more fundamental challenges in fuel mapping that need to be addressed first before accurate fuel maps are possible. As mentioned, we need to view fuels as biomass and understand those ecological processes and conditions that influence biomass properties over time and space. Once we understand fuel dynamics, we can then develop standardized sampling methods that describe fuels at their appropriate scales for quantifying reference conditions and select biophysical layers that represent those ecological processes that most influence fuel dynamics (Chap. 8). Spatial fuels databases containing all collected geo-referenced field data that is appropriately scaled to each fuel component can then be created so that spatially explicit fuels data can be accessible to everyone. Comprehensive, robust, and flexible fuel classifications can then be developed from these data (Chap. 7) that incorporate and account for the high variability in their design (Keane 2013). Categories in these new classifications can then be mapped using a fusion of the technologies mentioned here and any new technologies developed in the future. A new approach to fuels mapping is needed for enlightened fire management.

References

Abell CA (1937) Rate of spread and resistance to control data for Region 7 fuel types and their application to determine strength and speed of attack needed. Preliminary Report. U.S. Department of Agriculture, Forest Service, Asheville, pp 7

Agee JK, Skinner CN (2005) Basic principles of forest fuel reduction treatments. For Ecol Manage 211:83–96

Andersen H-E, McGaughey RJ, Reutebuch SE (2005) Estimating forest canopy fuel parameters using LIDAR data. Remote Sens Environ 94(4):441–449

Anderson HE (1982) Aids to determining fuel models for estimating fire behavior. USDA Forest Service Intermountain Research Station, General Technical Report INT-122 Ogden, Utah, USA, pp 22

Anderson M, Bourgeron P, Bryer MT, Crawford R, Engelking L, Faber-Langendoen D, Gallyoun M, Goodin K, Grossman DH, Landaal S, Metzler K, Patterson KD, Pyne M, Reid M, Sneddon L, Weakley AS (1998) International classification of ecological communities: terrestrial vegetation of the United States. The National Vegetation Classification System: list of types, vol. 2. The Nature Conservancy, Arlington, Virginia, pp 445

Arroyo LA, Pascual C, Manzanera JA (2008) Fire models and methods to map fuel types: the role of remote sensing. For Ecol Manage 256:1239–1252

Asner GP (1998) Biophysical and biochemical sources of variability in canopy reflectance. Remote Sens Environ 64(3):234–253

Ball GL, Guertin DP (1992) Advances in fire spread simulation. In: Proceedings on the third forest service remote sensing applications conference—Protecting natural resources with remote sensing. Tucson, Arizona USA, April 9–13 1992. American Society of Photogrammery and Remote Sensing, 5410 Grosvenor lane, Bethesda, Maryland USA, pp 241–249

Banks WG, Frayer HC (1966) Rate of forest fire spread and resistance to control in the fuel types of the eastern region. U.S. Department of Agriculture, Forest Service, Washington, DC. Fire Control Notes 27(2):10–13.

Barrett SW, Arno SF (1991) Classifying fire regimes and defining their topographic controls in the Selway-Bitterroot Wilderness. In: 11th Conference on Fire and Forest Meteorology, Missoula, MT, 1991, pp 233–245

Belfort W (1988) Controlled-scale aerial sampling photography: development and implications for multiresource inventory. J For 86(11):21–28

Bergen KM, Dobson MC (1999) Integration of remotely sensed radar imagery in modeling and mapping of forest biomass and net primary production. Ecol Model 122:257–274

Brandis K, Jacobson C (2003) Estimation of vegetative fuel loads using Landsat TM imagery in New South Wales, Australia. Internat J Wildland Fire 12:185–194

Brown AA, Davis K (1973) Forest fire control and use second edition. McGraw-Hill, New York

Burgan RE, Hardy CC (1994) Ground truthing a national AVHRR based vegetation fuels map. Proceedings of the 12th conference on fire and forest meteorology. Society of American Foresters, Bethesda, Maryland, pp 428–436

Burgan RE, Klaver RW, Klaver JM (1998) Fuel models and fire potential from satellite and surface observations. Int J Wildland Fire 8(3):159–170

Campbell J, Weinstein D, Finney M (1995) Forest fire behavior modeling integrating GIS and BEHAVE. USDA Forest Service, Washington

Chladil MA, Nunez M (1995) Assessing grassland moisture and biomass in Tasmania—The application of remote sensing and empirical models for a cloudy environment. Int J Wildland Fire 5(3):165–171

Chuvieco E, Salas J (1996) Mapping of spatial distribution of forest fire danger using GIS. Int J Geogr Inf Syst 10(3):333–345

De Vasconcelos MJP, Paul JCU, Silva S, Pereira JMC, Caetano MS, Catry FX, Oliveira TM (1998) Regional fuel mapping using a knowledge based system approach. 3rd International conference on forest fire research 14th Conference on fire and forest meteorology, November 1998, pp 2111–2123

Dixon R, Shipley W, Briggs A (1985) Landsat—a tool for mapping fuel types in the boreal forest of Manitoba: a pilot study. In: Proceedings of the Pecora X symposium, 1985. American Society for Photogrammetry and Remote Sensing, Falls Church, Virginia, pp 392–393

Elvidge CD (1988) Thermal infrared reflectance of dry plant materials: 2.5–20.0 micrometers. Remote Sens Environ 26:265–285

Erdody TL, Moskal LM (2010) Fusion of LiDAR and imagery for estimating forest canopy fuels. Remote Sens Environ 114(4):725–737. http://dx.doi.org/10.1016/j.rse.2009.11.002

Falkowski MJ, Gessler PE, Morgan P, Hudak AT, Smith AMS (2005) Characterizing and mapping forest fire fues using ASTER imagery and remote sensing. For Ecol Manage 217:129–146

Finney MA (1998) FARSITE: fire area simulator—model development and evaluation. United States Department of Agriculture, Forest Service Rocky Mountain Research Station, Research Paper RMRS-RP-4. Fort Collins, CO USA, pp 47

Finney MA (2005) The challenge of quantitative risk analysis for wildland fire. For Ecol Manage 211:97–108

Finney MA (2006) An overview of FlamMap fire modeling capabilities. In: Andrews PL, Butler BW (eds) Fuels management—how to measure success, Portland, OR, 28–30 March 2006. U.S. Department of Agriculture, Forest Service, Rocky Mountain Research Station, Proceedings RMRS-P-41 Fort Collins, CO USA, pp 213–220

Friedl MA, Michaelsen J, Davis FW, Walker H, Schimel DS (1994) Estimating grassland biomass and leaf area index using ground and satellite data. Int J Remote Sens 15(7):1401–1420

Gosz JR (1992) Gradient analysis of ecological change in time and space: implications for forest management. Ecol Appl 2(3):248–261

Grossman DH, Faber-Langendoen D, Weakley AS, Anderson M, Bourgeron P, Crawford R, Goodin K, Landaal S, Metzler K, Patterson K, Pyne M, Reid M, Sneddon L (1998) International classification of ecological communities: terrestrial vegetation of the United States Volume I. The National Vegetation Classification System: development, status, and applications., vol I. The Nature Conservancy, Arlington VA., USA

Habeck JR (1976) Forest, fuels and fire in the Selway-Bitterroot Wilderness, Idaho. 14th Tall timbers fire ecology conference 14:305–353

Hawkes B, Niemann O, Goodenough D, Lawson B, Thomson A, Sahle W, Fuglem P, Beck J, Bell B, Symingto P (1995) Forest fire fuel type mapping using GIS and remote sensing in British Columbia. In: Heit M, Parker HD, Shortreid A (eds) Proceedings of the symposium GIS applications in natural resources 2 – the 9th symposium on geographic information systems, Vancouver, British Columbia, Canada, pp 290–299

Hessburg PF, Reynolds K, Keane RE, James K, Salter RB (2010) Evaluating wildland fire danger and prioritizing vegetation and fuels treatments. In: Pye JM, Rauscher HM, Sands Y, Lee DC, Beatty JS (eds) Advances in threat assessment and their application to forest and rangeland management, Portland, OR, 2010. Gen. Tech. Rep. PNW-GTR-802. U.S. Department of Agriculture, Forest Service, Pacific Northwest and Southern Research Stations, pp 329–353

Heyerdahl EK, Brubaker LB, Agee JK (2001) Spatial controls of historical fire regimes: a multiscale example from the Interior West, USA. Ecology 82(3):660–678

Hornby LG (1935) Fuel type mapping in region one. J For 33(1):67–72

Hornby LG (1936) Fuel type mapping, its relation to forest fire control plannig. Northwest Sci 4:3–7

Huang S, Crabtree RL, Potter C, Gross P (2009) Estimating the quantity and quality of coarse woody debris in Yellowstone post-fire forest ecosystem from fusion of SAR and optical data. Remote Sens Environ 113(9):1926–1938. doi:10.1016/j.rse.2009.05.001

Jain A, Ravan SA, Singh RK, Das KK, Roy PS (1996) Forest fire risk modelling using remote sensing and geographic information systems. Curr Sci India 70(10):928–933

Jia GJ, Burke IC, Kaufmann MR, Goetz AFH, Kindel BC, Pu Y (2006) Estimates of forest canopy fuel attributes using hyperspectral data. For Ecol Manage 229:27–38

Karau EC, Keane RE (2010) Burn severity mapping using simulation modelling and satellite imagery. Int J Wildland Fire 19(6):710–724

Kasischke ES, French NFH, Bourgeau-Chavez LL, Ustin SL, Christensen NL (1998) Estimating the release of carbon from forest fires in Alaska using satellite remote sensing data. http://cstars.ucdavis.edu/papers/sustin/carbon/paper.html

Keane RE (2013) Describing wildland surface fuel loading for fire management: a review of approaches, methods and systems. Int J Wildland Fire 22(1):51–62

Keane RE, Reeves MC (2011) Use of expert knowledge to develop fuel maps for wildland fire management. In: Perera A, Drew A (eds) Expert knowledge and its application in landscape ecology. Island Press, New York, pp 211–228

Keane RE, Kevin CR, Running SW (1996) Simulating effects of fire on northern Rocky Mountain landscapes with the ecological process model Fire-BGC. Tree Physiol 16:319–331

Keane RE, McNicoll CH, Schmidt KM, Garner JL (1997) Spatially explicit ecological inventories for ecosystem management planning using gradient modeling and remote sensing. In: Proceedings of the sixth forest service remote sensing applications conference—remote sensing; people in partnership with technology. American Society of Photogrammetry and Remote Sensing. Bethesda, MD., USA, pp 135–146

Keane RE, Garner JL, Schmidt KM, Long DG, Menakis JP, Finney MA (1998a) Development of input spatial data layers for the FARSITE fire growth model for the Selway-Bitterroot Wilderness complex, USA. USDA Forest Service, Rocky Mountain Research Station, General Technical Report RMRS-GTR-3 Fort Collins, CO USA, pp 66

Keane RE, Long DG, Schmidt KM, Mincemoyer SA, Garner JL (1998b) Mapping fuels for spatial fire simulations using remote sensing and biophysical modeling. Natural resource management using remote sensing and GIS: proceedings of the seventh forest service remote sensing applications conference, Nassau Bay, Texas, April 6–10, 1998. American Society for Photogrammetry and Remote Sensing c1998, Bethesda

Keane RE, Mincemoyer SA, Schmidt KM, Menakis JP, Long DG, Garner JL (2000) Mapping vegetation and fuels for fire management on the Gila National Forest Complex. USDA Forest Service, Rocky Mountain Research Station General Technical Report RMRS-GTR-46-CD Fort Collins, CO USA, pp 67

Keane RE, Burgan RE, Wagtendonk JV (2001) Mapping wildland fuels for fire management across multiple scales: integrating remote sensing, GIS, and biophysical modeling. Int J Wildland Fire 10:301–319

Keane RE, Cecilia McNicoll, Rollins MG (2002) Integrating ecosystem sampling, gradient modeling, remote sensing, and ecosystem simulation to create spatially explicit landscape inventories. USDA Forest Service, Rocky Mountain Research Station, General Technical Report RMRS-GTR-92 Fort Collins, CO USA, pp 61

Keane RE, Frescino TL, Reeves MC, Long J (2006) Mapping wildland fuels across large regions for the LANDFIRE prototype project. In: Rollins MG, Frame C (eds) The LANDFIRE prototype project: nationally consistent and locally relevant geospatial data for wildland fire management, 2006. USDA Forest Service Rocky Mountain Research Station, General Technical Report RMRS-GTR-175 Ogden, UT USA, pp 367–396

Keane RE, Holsinger L, Pratt S (2006b) Simulating historical landscape dynamics using the landscape fire succession model LANDSUM version 4.0. USDA Forest Service Rocky Mountain Research Station, General Technical Report RMRS-GTR-171CD Fort Collins, CO USA, pp 73

Keane RE, Rollins M, Zhu ZL (2007) Using simulated historical time series to prioritize fuel treatments on landscapes across the United States: the LANDFIRE prototype project. Ecol Model 204(3–4):485–502. doi:10.1016/j.ecolmodel.2007.2.005

Keane RE, Loehman RA, Holsinger LM (2011) The FireBGCv2 landscape fire and succession model: a research simulation platform for exploring fire and vegetation dynamics. U.S. Department of Agriculture, Forest Service, Rocky Mountain Research Station, General Technical Report RMRS-GTR-255 Fort Collins, CO USA, pp 137

Keane R, Gray K, Bacciu V, Leirfallom S (2012a) Spatial scaling of wildland fuels for six forest and rangeland ecosystems of the northern Rocky Mountains, USA. Landsc Ecol 27(8):1213–1234. doi:10.1007/s10980-012-9773–9

Keane RE, Gray K, Bacciu V (2012b) Spatial variability of wildland fuel characteristics in northern Rocky Mountain ecosystems. USDA Forest Service Rocky Mountain Research Station, Research Paper RMRS-RP-98 Fort Collins, Colorado USA, pp 58

Keane RE, Herynk JM, Toney C, Urbanski SP, Lutes DC, Ottmar RD (2013) Evaluating the performance and mapping of three fuel classification systems using forest inventory and analysis surface fuel measurements. For Ecol Manage 305:248–263. http://dx.doi.org/10.1016/j.foreco.2013.06.001

Keramitsoglou I, Kontoes C, Sykioti O, Sifakis N, Xofis P (2008) Reliable, accurate, and timely forest mapping for wildfire management using ASTER and Hyperion satellite imagery. For Ecol Manage 255:3556–3562

Kessell SR (1976) Gradient modeling: a new approach to fire modeling and wilderness resource management. Environ Manage 1(1):39–48

Kourtz PH (1977) An application of Landsat digital technology to forest fire fuel mapping. In: Dube DE (ed) Fire ecology in resource mangement: a workshop. Northern Forest Research Centre, Information Report NOR-X-210 Edmonton, Alberta, pp 79–81

Krasnow K (2007) Forest fuel mapping and strategic wildfire mitigation in the Montane zone of Boulder County, Colorado. Graduate Thesis, University of Colorado, Boulder, CO USA, pp 55

Krivtsov V, Vigy O, Legg C, Curt T, Rigolot E, Lecomte I, Jappiot M, Lampin-Maillet C, Fernandes P, Pezzatti GB (2009) Fuel modelling in terrestrial ecosystems: an overview in the context of the development of an object-orientated database for wild fire analysis. Ecol Model 220(21):2915–2926

Lanorte A, Lasaponara R, Murgante B, Gervasi O, Iglesias A, Taniar D, Apduhan B (2011) On the use of satellite remote sensing data to characterize and map fuel types. In: Computational science and its applications—ICCSA 2011, vol 6783. Lecture Notes in Computer Science. Springer Berlin/Heidelberg, pp 344–353. doi:10.1007/978-3-642-21887-3_27

Lasaponara R, Lanorte A (2007a) On the capability of satellite VHR Quickbird data for fuel type characterization in fragmented landscapes. Ecol Model 204:79–84

Lasaponara R, Lanorte A (2007b) Remotely sensed characterization of forest fuel types by using satellite ASTER data. Int J Appl Earth Obs Geoinf 9(3):225–234. http://dx.doi.org/10.1016/j.jag.2006.08.001

Leenhouts B (1998) Assessment of biomass burning in the conterminous United States. Conserv Ecol 2(1):1–23

Lenihan JM, Daly C, Bachelet D, Neilson RP (1998) Simulating broad scale fire severity in a dynamic global vegetation model. Northwest Sci 72:91–103

Loveland TR, Merchant JM, Ohlen DO, Brown JF (1991) Development of a land-cover characteristics database for the conterminous U.S. Photogramm Eng Remote Sens 57(11):1453–1463

Loveland TR, Ohlen DO, Brown JF, Reed BC, Merchant JW (1993) Prototype 1990 conterminous United States land cover characteristics data set EROS Data Center, US Geological Survey, CD-ROM 9307 Souix Falls, SD USA

McKenzie D, Raymond CL, Kellogg L-KB, Norheim RA, Andreu A, Bayard AC, Kopper KE, Elman E (2007) Mapping fuels at multiple scales: landscape application of the fuel characteristic classification system. Can J For Res 37:2421–2437. doi:10.1139/x07-056

Menakis JP, Keane RE., Long DG (2000) Mapping ecological attributes using an integrated vegetation classification system approach. J Sustain For 11(1–2):245–265

Merrill EH, Bramble-Brodahl MK, Marrs RW, Boyce MS (1993) Estimation of green herbaceous phytomass from Landsat MSS data in Yellowstone National Park. J Rang Manage 46:151–157

Miller JD, Danzer SR, Watts JM, Stone S, Yool SR (2003) Cluster analysis of structural stage classes to map wildland fuels in a Madrean ecosystem. J Environ Manage 68(3):239–252

Millington AC, Critchley RW, Douglas TD, Ryan P (1994) Estimating woody biomass in sub-Saharan Africa. The World Bank, Washington, DC

Morris WG (1970) Photo inventory of fine logging slash. Photogramm Eng Remote Sens 36:1252–1256

Muraro SJ (1970) Slash fuel inventories from 70 mm low-level photography. Canadian Forest Service, Information Report 1268, Ottawa, Ontario, Canada, pp 63

Mutlu M, Popescu SC, Stripling C, Spencer T (2008) Mapping surface fuel models using lidar and multispectral data fusion for fire behavior. Remote Sens Environ 112(1):274–285. http://dx.doi.org/10.1016/j.rse.2007.5.005

New Jersey Department of Conservation and Development U.S. Department of Agriculture FS (1942) Forest fuel types of New Jersey. New Jersey Department of Conservation and Development, Trenton, NJ

Ohmann JL, Spies TA (1998) Regional gradient analysis and spatial pattern of woody plant communities of Oregon forests. Ecol Monogr 68(2):151–182

Omernik JM (1987) Ecoregions of the conterminous United States. Annu Assoc Am Geogr 77(1):118–125

Oswald BP, Fancher JT, Kulhavy DL, Reeves HC (1999) Classifying fuels with aerial photography in East Texas. Int J Wildland Fire 9(2):109–113

Ottmar RD, Vihnanek R, Alvarado E (1994) Forest health assessment: air quality tradeoffs. In: Proceedings of the 12th international conference on fire and forest meterology, Jekyll Island, Georgia, USA, October 26–28, 1993. Society of American Foresters, Bethesda, MD, USA, pp 47–62

Pala S, Taylor D, Holder G (1990) Integrating satellite-derived forest fuel data into fire management decision support models. In: Proceedings, second national GIS conference, Ottawa, Canada, March 5–8, 1990

Peterson SH, Franklin J, Roberts DA, van Wagtendonk JW (2012) Mapping fuels in Yosemite National Park. Can J For Res 43(1):7–17. doi:10.1139/cjfr-2012-0213

Pierce AD, Farris CA, Taylor AH (2012) Use of random forests for modeling and mapping forest canopy fuels for fire behavior analysis in Lassen Volcanic National Park, California, USA. For Ecol Manage 279:77–89. doi:http://dx.doi.org/10.1016/j.foreco.2012.05.010

Quigley TM, Graham RT, Haynes RW (1996) An integrated scientific assessment for ecosystem management in the Interior Columbia River Basin and portions of the Klamath and Great Basins. US Department of Agriculture, Forest Service, Pacific Northwest Research Station, PNW-GTR-382 Portland, OR., USA, pp 126

Reeves MC, Ryan KC, Rollins MC, Thompson TG (2009) Spatial fuel data products of the LANDFIRE project. Int J Wildland Fire 18:250–267

Reinhardt E, Lutes D, Scott J (2006) FuelCalc: a method for estimating fuel characteristics. In: Andrews PL, Butler BW (eds) Fuels management—how to measure success, Portland, OR, 2006a. I U.S. Department of Agriculture, Forest Service, Rocky Mountain Research Station, Proceedings RMRS-P-41 Fort Collins, CO, pp 273–287

Riaño D, Meier E, Allgöwer B, Chuvieco E, Ustin SL (2003) Modeling airborne laser scanning data for the spatial generation of critical forest parameters in fire behavior modeling. Remote Sens Environ 86(2):177–186

Rignot E, Way J, Williams C, Viereck L (1994) Radar estimates of aboveground biomass in boreal forests of interior Alaska. IEEE Trans Geosci Remote Sens 32(5):1117–1124

Rollins MG (2009) LANDFIRE: a nationally consistent vegetation, wildland fire, and fuel assessment. Int J Wildland Fire 18(3):235–249. doi:10.1071/WF08088

Rollins MG, Keane RE, Parsons R (2004) Mapping ecological attributes using gradient analysis and remote sensing. Ecol Appl 14(1):75–95

Root RR, Wagtendonk J (1999) Hyperspectral analysis of multi-temporal Landsat TM data for mapping fuels in Yosemite National Park. In: Annual ASPRS Conference, Portland, OR., USA, May 17–21, 1999. Society of American Photogrammeters, Bethesda, MD., USA, pp 399–408

Root RR, Stitt SCF, Nyquist MO, Waggoner GS, Agee JK (1985) Vegetation and fire fuel models mapping of North Cascades National Park. In: The tenth William T. Pecora remote sensing symposium, Fort Collins, CO, 1985, pp 287–294

Saatchi S, Halligan K, Despain DG, Crabtree RL (2007) Estimation of forest fuel load from radar remote sensing. Geosci Remote Sens, IEEE Trans 45(6):1726–1740. doi:10.1109/tgrs.2006.887002

Salas J, Chuvieco E (1994) Geographic information systems for wildland fire risk mapping. Wildfire 3(2):7–13

Schmidt KM, Menakis JP, Hardy CC, Hann WJ, Bunnell DL (2002) Development of coarse-scale spatial data for wildland fire and fuel management. USDA Forest Service Rocky Mountain Research Station, General Technical Report RMRS-GTR-87 Fort Collins, CO, USA, pp 41

Scott J, Burgan RE (2005) A new set of standard fire behavior fuel models for use with Rothermel's surface fire spread model. USDA Forest Service Rocky Mountain Research Station, General Technical Report RMRS-GTR-153. Fort Collins, CO USA, pp 66

Sikkink PG, Keane RE (2008) A comparison of five sampling techniques to estimate surface fuel loading in montane forests. Int J Wildland Fire 17(3):363–379. doi:Doi 10.1071/Wf07003

Skidmore AK (1989) A comparison of techniques for calculating gradient and aspect from a gridded digital elevation model. Int J Geogr Info Sys 3(4):323–334

Thornton PE, Law BE, Gholz HL, Clark KL, Falge E, Ellsworth DS, Goldstein AH, Monson RK, Hollinger DY, Falk M, Chen J, Sparks JP (2002) Modeling and measuring the effects of disturbance history and climate on carbon and water budgets in evergreen needleleaf forests. Agric For Meteorol 113:185–222

Varga TA, Asner GP (2008) Hyperspectral and LiDAR remote sensing of fire fuels in Hawaii Volcanoes National Park. Ecol Appl 18(3):613–623. doi:10.1890/07-1280.1

Vasconcelos JJPd, Paul JCU, Silva S, Pereira JMC, Caetono MS, Catry FX, Oliveira TM (1998) Regional fuel mapping using a knowledge based system approach. In: Proceedings: III international conference on forest fire research and 14th conference on fire and forest meterology, vol. II, 1998. pp 2111–2123

Verbyla DL (1995) Satellite remote sensing of natural resources. Lewis Publishers, CRC Press Valley Stream, New York

Vogelmann JE, Kost JR, Tolk B, Howard S, Short K, Xuexia C, Chengquan H, Pabst K, Rollins MG (2011) Monitoring landscape change for LANDFIRE using multi-temporal satellite imagery and Ancillary data. Selected topics in applied earth observations and remote sensing. IEEE J 4(2):252–264. doi:10.1109/jstars.2010.2044478

Willis JM (1985) Applications of Landsat imagery for fire fuels mapping projects covering large geographic areas. In: Proceedings of the Pecora X symposium, 1985. American Society for Photogrammetry and Remote Sensing, Falls Church, Virginia, USA, pp 394–395

Wilson BA, Ow CFY, Heathcott M, Milne D, McCaffrey TM, Ghitter G, Franklin SE (1994) Landsat MSS classification of fire fuel types in Wood Buffalo National Park, Northern Canada. Glob Ecol Biogeogr Lett 4:33–39

Xiao-rui T, McRae D, Li-fu S, Ming-yu W (2005) Fuel classification and mapping from satellite imagines. J For Res 16(4):311–316. doi:10.1007/bf02858198

Chapter 10
Fuel Concepts

Sometimes a concept is baffling not because it is profound but because it is wrong
E.O. Wilson, American ecologist

10.1 Introduction

As detailed in this book, wildland fuels are defined, described, and quantified based on their use in fire behavior and effects applications, which often ignore the great biological complexity and rich ecological context needed to fully understand them. Woody fuels, for example, are separated into size classes based on rate of moisture loss (Fosberg 1970) rather than based on rates of decomposition or deposition. The fire behavior context may also limit or compromise fuel management concepts, applications and actions. This chapter presents a set of common fuel concepts used in fire science and management that need further qualification if they are used correctly in an ecological context.

10.2 Important Fuel Concepts

10.2.1 Flammability

Flammability is not a useful concept without considering a myriad of combustion qualities and their distribution across multiple spatial and temporal scales. Gisborne (1947) noted that most fuels are similar chemically and the major differences in burning of different kinds of wood, brush, or grass are mainly due to variations in physical properties rather than chemical composition. Differences in how fuels burn are typically referred to *flammability*. The concept of flammability is interpreted differently all over the world. It is often defined as the relative ease at which fuels ignite but the causal mechanisms of flammability are

R. E. Keane, *Wildland Fuel Fundamentals and Applications*,
DOI 10.1007/978-3-319-09015-3_10

interpreted differently. Anderson (1966) characterized fuel flammability by three criteria:

1. Ignitability. Ability to ignite at the lowest temperature with the least energy input
2. Sustainability. Ability to continue flaming combustion when heat source is removed
3. Combustibility. Ability to facilitate combustion to have short burn times and the highest proportion of consumption

These characteristics can be evaluated at particle, component, layer, and fuelbed scales. However, the metrics used to rate these criteria can rarely be measured in absolute terms (Mak 1988). Most assessments inevitably use the components of combustion to rate flammability (Dimitrakopoulos and Papaioannou 2001), such as measuring time to ignition at a standard temperature or heat flux. Dimitrakopoulos (2001), however, rated flammability based on five fuel characteristics including heat content, ash content, silica-free ash content, surface area-to-volume ratios and particle density. Liodakis et al. (2002) found that flammability (ease of ignition) was insensitive to the inorganic concentration of the fuel, but rather was closely tied to rates of cellulose decomposition. Dimitrakopoulos and Papaioannou (2001) found moisture content was the overriding factor dictating the flammability of Mediterranean fuels. Many believe that the chemical compositions of live and some dead fuels are important to flammability because of the presence of resins, crude fats, oils, and other volatile substances might facilitate ignition and consumption (Jolly et al. 2012; Philpot 1969). In most cases, however, fuel differs in flammability because of physical, morphological, and ecophysiological characteristics, and less because of chemical characteristics. For example, some species (e.g., sagebrush, saltbrush) have ecophysiological adaptations to allow them to survive when their foliar moisture is low (<50 %) thereby making them more flammable. Because flammability cannot be objectively assessed using standardized tests, the concept is best used in a relative, qualitative sense.

The problem with flammability is that it must be interpreted across many spatial and temporal scales and across many combustion qualities to be useful for fire management. Some shrub species may be highly flammable because of small, dense foliage, but their foliage may be flammable for only a small portion of the year because of high live fuel moistures. Grasses are highly flammable when cured, but if they are not connected in space, or if they exist in communities that contain abundant inflammable species, then their flammability at broader stand scales may be lower. Some fuel types, such as resin-saturated wood, may have higher heat contents contributing to high flammability, but they may also have a low SAVR that serves to decrease flammability. Rotten wood has low particle densities that may increase the potential for ignition at higher moistures over sound wood, but rotten particles also have lower heat contents, that might decrease flammability at the particle level, and the rotten particles may be widely scattered across the landscape thereby decreasing flammability at coarser spatial scales.

A more current example of the challenges involved in evaluating flammability concerns the extensive areas across North America that have been killed by pine beetles (Jenkins et al. 2012). The dead, red foliage in pine stands killed by the bark beetles appear highly flammable and studies show that the suspended dead foliage is generally more flammable than green needles (Jolly et al. 2012). However, these red needles fall to the ground in 1–3 years, leaving a stand of dead trees with significantly reduced crown fire potential. Further, beetles rarely kill all trees in a stand; patchworks of 10–90 % beetle mortality in outbreak stands are common (Schoennagel et al. 2012). The green pine needles may be just as flammable as red needles at the height of an extreme fire season, especially in high wind. Without spatial and temporal context, it will be always difficult to rate the flammability of fuels.

10.2.2 Fuel Treatments

Management activities aimed at reducing fuels should always consider ecology first to increase treatment effectiveness in the long run. A fuel treatment is defined as the manipulation, modification, and/or removal of wildland fuels to reduce the likelihood of ignition, to reduce potential fire intensity and spread rates, and/or to lessen potential damage and resistance to control (NWCG 2006). However, Reinhardt et al. (2008) caution that the primary purpose of fuel treatments should rarely be to reduce burned area or limit fire growth. Rather, fuel treatments should be implemented to reduce fire intensity and severity, so that if an area burns, the fire will have minimal impacts on the ecosystem and it will have a lesser chance of harming people and property. Fuel treatments should also allow firefighters to safely and effectively fight fire to prevent the loss of property or life, even though fighting the fire will probably increase future suppression efforts. So the question facing fire managers is how to treat fuels to accomplish these multiple and sometimes conflicting objectives.

The problem in designing fuel treatments that don't consider wildland fuel ecology is that those treatments will probably be ineffective and even counterproductive in the long run. Fuel treatments should also restore ecosystems to make them effective at both reducing fuels and useful for other land management objectives. Implementing treatments for the sole purpose of reducing fuels ignores decades of research on the benefits of an ecosystem management approach (Crow and Gustafson 1997). It makes little sense to modify fuels with a treatment that creates unconventional stand structures that will quickly compromise the efficacy of the treatment and will contribute to declines in ecosystem health. Many mechanical canopy fuel treatments, for example, remove trees to reduce canopy bulk density (CBD) and increase canopy base height (CBH) without regard to the shade tolerance and fire adaptations of the residual species. Leaving shade-tolerant, fire-sensitive species may create stand structures that are highly susceptible to future fire and have low resilience to other disturbances (Agee and Skinner 2005). Moreover, shade-tolerant residual species can quickly regenerate in treated areas thereby rapidly lowering CBH and quickly rendering the primary treatment ineffective. When the residual stand inevitably does burn, it may have a higher probability of experiencing extensive fire-caused mortality because it is composed of fire susceptible species.

Mastication is a mechanical treatment that uses specialized equipment to shred, chip, or break apart dead and live trees to raise CBH, decrease CBD, and reduce fuelbed depth. This treatment is amenable to many managers because it is relatively cheap and poses little risk when compared with prescribed fire. However, the tremendous amount of shattered fuel left on the forest floor may cause adverse ecological consequences to the treated stand, such as increased rainfall interception, reduced plant regeneration, altered nutrient cycling, and increased soil insulation (Kane et al. 2010). Moreover, plant mortality may be high when masticated stands eventually burn in wildfires because of deep soil heating from smoldering combustion of the thick duff and litter layers and high fire intensities from burning the dense surface fuels (Kreye et al. 2014). Short-term decreases in fire hazard may be eventually overwhelmed by long-term declines in ecological integrity for those fuel treatments that don't address ecological concerns.

Current efforts in the restoration of fire-prone ecosystems often design treatments to create stands that, when burned by wildfire, will experience fire effects that were common historically and that will leave the landscape with much of the prefire character. Restoration actions in frequent-fire ecosystems, such as ponderosa pine forests, often retain large, relic individuals that are able to survive future wildfires (Agee and Skinner 2005). Retaining large trees that have survived many disturbances and have lived through many climate fluctuations will ensure high survival rates for future disturbances and offer the best genetic sources for the future. Ecosystems with long fire return intervals, such as the Canadian boreal or the Rocky Mountain lodgepole pine forests, experience primarily high-intensity crown fires so the goal of keeping large trees may not be a concern. Instead of emphasizing reducing fuels, an approach that balances what can survive a wildfire with what fire behaviors and effects are acceptable both ecologically and socially will often result in both fuel reduction and beneficial ecological restoration.

Another important factor in designing fuel treatments is the condition of the vegetation being treated. Most fire-prone ecosystems need fuel treatments because fire has been excluded from them for decades. As a result, plants are highly stressed because competition for light, water, and nutrients has been amplified due to both increased number of individuals and high surface fuel buildups that may have altered water and nutrient cycling (Chap. 6). Moreover, trees that haven't experienced fire in the past are not well equipped to survive fire in the future. Trees that have been burned with low-severity fire, for example, often develop resin ducts needed to compartmentalize injury and survive fires in the future. And because of competition for soil moisture, trees tend to concentrate fine roots in the thick duff and litter layer that has accumulated in the absence of fire, so when fires do occur in these fire-excluded stand, trees may experience high root and plant mortality. Fuel treatments that fail to recognize initial stand health in their design may cause additional mortality that result in even greater posttreatment surface fuel loadings and unsustainable ecosystems.

Obviously, the best fuel treatment option for both reducing fuels and restoring fire-prone ecosystems is fire implemented as prescribed burns or controlled wildfires (allowing wildfires to burn). Fire has shaped past landscapes so it can best be used tomorrow for restoring the fire-excluded ecosystems of today. It performs many attractive actions such as (1) reducing canopy fuels by killing fire-susceptible trees and dense regeneration, (2) creating heterogeneous patches and landscapes, (3) consuming fuels, (4) facilitating nutrient and water cycling, and (5) favoring fire-adapted vegetation leading to more resilient landscapes. The dilemma facing most fire managers, however, is how to reintroduce fire into areas that now have such great buildups of canopy and surface fuels that any fire may kill large trees and cause uncharacteristically high plant mortality. Burning fire-excluded areas with prescribed fire is difficult because weather conditions that are needed to achieve targeted fire intensities may be rare and the risk of damaging ecosystems, burning property, or harming people may be great. The successful reintroduction of fire will take multiple treatments staggered in time. In the first entry, mechanical treatments may be needed to reduce canopy fuels so that fire treatments can follow to consume surface fuels and lower residual tree regeneration without high tree mortality (North et al. 2012). Further, it will take several entries into treated stands over time to obtain the stand structures that will be resilient in the face of climate change and resistant to future disturbance events. Some estimate it will take three to seven burns to return landscapes to a semblance of the historic past. Irrespective of treatment strategy, wildfires will eventually burn most areas regardless of the level of fire suppression so designing effective treatment regimes that protect and restore ecosystems while reducing fuels will need to balance society's tolerance for fire with an ecosystems' ability to accumulate fuels using enlightened fire management that fully integrates mechanical treatments and prescribed fire with the eventual wildfires.

10.2.3 Fuel Treatment Longevity

Treatment longevity depends on the highly variable fire and fuel characteristics being used to evaluate it and has little meaning without considering space and time. Since most fuel treatments are costly, it is important to know how long they will last before another one is needed. The time over which fuel treatments are effective is often called treatment *longevity*. Longevity is important for planning and scheduling future fuel treatments, but, similar to flammability, assessing treatment longevity is difficult because of the temporal and spatial complexities of fuelbed dynamics. The wide ranges of measurements that can be used to assess longevity coupled with the incredible variability and complexity of fuelbeds as they interact with the biophysical environment make estimating longevity nearly impossible.

The most important challenge in quantifying longevity is deciding which suite of fire and fuel variables to use to assess how long treatments are effective. An objective of a fuel treatment, for example, might be to reduce surface fuel loadings or decrease CBD below a threshold value, or to create fuelbed conditions, that when burned, exhibited fire behavior below some threshold parameter. If fuel loadings are used as measures of longevity, then the length of time the treated stand stays below a threshold fuel loading might be used as a measure of longevity. Alternatively, if fire behavior attributes are used, then the measured fuel conditions can be input to fire behavior models, such as BEHAVE or NEXUS, to compute fire behavior metrics, such as rate of spread, fire intensity, and torching index, that are then used to estimate longevity by comparing against a corresponding threshold value.

The problem is that countless fire behavior and fuel characteristics can be used to evaluate longevity, and among them they have different (1) rates of change over time, (2) threshold and importance values to management, (3) degrees of correlation with each other, and (4) spatial distributions. Commonly used canopy fuel treatments involve removing trees (reducing CBD and increasing CBH), but this may open the stand to increases in incipient herbaceous vegetation resulting in a flashy surface fuel layer that has a higher rate of spread than the previous fuelbed (Kane et al. 2010). An example of this dilemma concerns the treatment of ponderosa pine ecosystems that historically experienced frequent fires, but as a result of fire exclusion, these forests are now densely populated with trees (high CBD, low CBH) that heavily shade the forest floor making it mostly devoid of shrub and herbaceous fuels with thick duff and litter layers and high accumulations of woody fuels. If spread rate were used as a measure, then the spread rates before the treatment might actually be lower than the spread rates in the herbaceous-dominated posttreatment stand. In other treatments, tree removal may result in slash that increases surface woody fuel loadings creating conditions where decreases in crown fire potential from CBH reductions may be offset by the increases in surface fire intensities and concomitant deeper soil heating (Pollet and Omi 2002). Moreover, the use of multiple fire behavior and fuels measures to evaluate longevity might produce contradictory results. Values of one fire or fuel metric might always exceed threshold values even though the values of other fire and fuel metrics remain under their threshold values. Some surface fuel threshold values might be so high, for example, that a target low torching index might be impossible.

A more holistic approach is needed to replace the notion of treatment longevity with a broader view of fuel in a landscape and successional ecology framework. First, longevity, by definition, demands a consideration of scale and context. Most managers and scientists tend to evaluate longevity at the stand scale because this is the scale of treatment. However, to be fully effective, fuel treatments need to be implemented at landscape scales to take advantage of adjacency, contagion, and connectivity in fuel patch heterogeneity. It makes little sense to only treat a 10-ha ponderosa stand, for example, when it is surrounded by a 10,000-ha dense forest with heavy surface fuels. The treated ponderosa pine stand might be effective longer if it is surrounded by other treated stands. And since fuel deposition rates are dictated by vegetation development, a consideration of the vegetation species and structure before and after the treatment is critical to ensure that treatments are appropriate for the biophysical conditions so that they are effective longer.

10.2.4 Hazard and Risk

Hardy (2005) elucidated three major problems with using hazard and risk in fire science and management. First, there are too many definitions of fire hazard and risk that make consistent quantification difficult in fire management. An assessment of hazard using the NWCG (2006) definition (condition that can cause injury, illness or death of personnel, or damage to, or loss of equipment or property) will be entirely different than one using the definition proposed by Bachmann and Allgower (2001) (the potential fire behavior for a fuel type, regardless of the fuel type's weather-influenced fuel moisture content). Next, assessments of hazard and risk demand a temporal and spatial context, which has been missing from many past efforts. Any evaluation must include the extent and pattern of the hazard and how long it remains a hazard both within the fire season and over the long term. And last, the context of hazard and risk must include an assessment of the historical or "natural" fire regime and the ecosystem being evaluated. Chaparral ecosystems, for example, may have high hazard for fire management but these shrublands may only burn in high-intensity fires. Without an ecological context, ecosystems that historically experienced large, infrequent, and high-intensity fires will always be considered hazardous with high risk.

The quantification and mapping of hazard and risk should recognize that fuels are always changing in space and time (Chap. 6). Too often, hazard and risk projections fail to integrate how long fuels on a stand or landscape remain hazardous or when less hazardous fuel complexes become more hazardous. Fuel maps used as input in fire hazard and risk software packages, such as FLAMMAP (Finney 2006), seldom address changing fuel conditions over time. In the beetle example mentioned above, the elevated fire hazard because of red needles on pine trees killed by mountain pine beetles only exists until the needles fall off the trees (1–3 years). If these red-needle pines are in environments where fire seasons are short and infrequent, such as lodgepole and whitebark pine forests, there is a low chance that the surface fuels will be dry enough to support fire while the red needles remain on the trees. Moreover, the current fire behavior fuel models (FBFMs) used in most fire hazard and risk assessments may be too coarse to detect subtle but major changes in fuel loadings over time or after disturbances (Chap. 7). A fully integrated hazard and risk assessment must simulate ecosystem change to meld dynamic fuel maps with all possible weather and climate scenarios in both a fire behavior and fire effects context (Keane and Finney 2003; Finney 2005).

10.3 A Fuel Ecology Approach

It will always be difficult to thoroughly understand wildland fuels if they are exclusively analyzed and studied through the lens of fire behavior. When fuels are defined, stratified, classified, and mapped for combustion science applications, the increased uncertainty resulting from inappropriate stratifications may overwhelm most fire analyses producing misleading and inappropriate results. Stratifying fuels into broad life forms or aggregating all logs into one class regardless of size may mask differences between costly treatments that may be important for fuel management. More comprehensive, accurate, and useful fuel applications and products will hopefully be developed to meet the needs of fire management by expanding the focus of wildland fuels beyond the fire behavior construct to a more ecological paradigm. A fuel ecology approach can meld important ecological relationships with the needs of fire science to continually create novel, insightful, and useful future fire applications for the future.

An interesting dilemma is that, if an ecological approach is important for describing fuels for fire applications, then why have the current fuel inputs to fire models continued to satisfy fire managers and researchers? Predictions from these models are used extensively in fire management with acceptable results yet none deal with spatial variability, ecologically inappropriate fuel components, and temporal dynamics detailed in this book. A partial answer might be that weather and topography, not fuels, are more important for fire behavior predictions under most wildfire conditions (Bessie and Johnson 1995); strong winds, dry air, and high temperatures may exert more influence on fire behavior than the fuel complex. Another reason may be that the models used in fire management are one-dimensional point models that are designed to be used for small, homogeneous areas and the high extrapolation error of point estimates across space may exceed the variability of fuel characteristics. It could also be that, since the FBFMs used in US fire behavior prediction systems are calibrated or adjusted to compute behavior values that match observed fire behavior (Burgan and Rothermal 1984), this calibration, while entirely subjective, has indirectly accounted for the ecological inconsistencies and variability in the classification. Finally, it might be that the high uncertainty in fuel sampling and fire behavior measurements make it difficult to actually validate the fire behavior and effects predictions (Chap. 7). One thing is certain, if fire managers and researchers want more accurate and consistent estimates of fire, future fire models must account for fuel landscape ecology to comprehensively simulate fire behavior and effects (Thaxton and Platt 2006; Parsons et al. 2010).

The benefits of a fuel ecology approach are numerous. Fuel properties, especially loading and density, can be given a better temporal context by linking the ecological processes of phenology, decomposition, disturbance, and deposition to ecologically based fuel components. More appropriate and accurate sampling techniques can be devised if fuel types and components are described using ecological relationships rather than fire modeling parameterizations. Fuelbeds can be more comprehensively described if the classifications emphasize both ecological and fire behavior aspects. The mapping of fuel characteristics can be improved by studying the landscape ecology of fuels and the processes that control them. Inputs to newer and better fire behavior models will be easier to collect if an ecological context is used to design input data structures. While the current fuel description systems have great value to fire management, the future of fire science is best served if new fire science applications are developed with a focus on both ecological and combustion science.

References

Agee JK, Skinner CN (2005) Basic principles of forest fuel reduction treatments. For Ecol Manage 211:83–96

Anderson HE (1966) Flammability features of forest fuels. Unpublished Report on file at the Missoula Fire Sciences Laboratory, Missoula, MT 22 pp

Bachmann A, Allgower B (2001) A consistent wildland fire risk terminology is needed! Fire Manage Today 61(4):28–33

Bessie WC, Johnson EA (1995) The relative importance of fuels and weather on fire behaviour in subalpine forests. Ecology 76(3):747–762

Burgan RE, Rothermal RC (1984) BEHAVE: fire behavior prediction and fuel modeling system–FUEL subsystem. USDA Forest Service

Crow TR, Gustafson EJ (1997) Ecosystem management: managing natural resources in time and space. In: Kohm KA, Franklin JF (eds) Creating forestry for the 21st century. Island Press, Washington, DC, pp 215–229

Dimitrakopoulos AP (2001) A statistical classification of Mediterranean species based on their flammability components. Int J Wildland Fire 10(2):113–118. doi:10.1071/WF01004

Dimitrakopoulos AP, Papaioannou K (2001) Flammability assessment of Mediterranean forest fuels. Fire Technol 37(2):143–152. doi:10.1023/a:1011641601076

Finney MA (2005) The challenge of quantitative risk analysis for wildland fire. For Ecol Manage 211:97–108

Finney MA (2006) An overview of FlamMap fire modeling capabilities. In: Andrews PL, Butler BW (eds) Fuels management—how to measure success, Portland, OR, 28–30 March 2006. U.S. Department of Agriculture, Forest Service, Rocky Mountain Research Station, Proceedings RMRS-P-41, pp 213–220

Fosberg MA (1970) Drying rates of heartwood below fiber saturation. For Sci 16:57–63

Gisborne HT (1947) Fundamentals of fire behavior. Fire Control Notes 9(1):13–24

Hardy CC (2005) Wildland fire hazard and risk: problems, definitions, and context. For Ecol Manage 211:73–82

Jenkins MJ, Page WG, Hebertson EG, Alexander ME (2012) Fuels and fire behavior dynamics in bark beetle-attacked forests in Western North America and implications for fire management. For Ecol Manage 275:23–34. doi:http://dx.doi.org/10.1016/j.foreco.2012.02.036

Jolly WM, Parsons RA, Hadlow AM, Cohn GM, McAllister SS, Popp JB, Hubbard RM, Negron JF (2012) Relationships between moisture, chemistry, and ignition of Pinus contorta needles during the early stages of mountain pine beetle attack. For Ecol Manage 269:52–59. doi: http://dx.doi.org/10.1016/j.foreco.2011.12.022

Kane JM, Varner JM, Knapp EE, Powers RF (2010) Understory vegetation response to mechanical mastication and other fuels treatments in a ponderosa pine forest. Appl Veg Sci 13(2):207–220. doi:10.1111/j.1654-109X.2009.01062.x

Keane RE, Finney MA (2003) The simulation of landscape fire, climate, and ecosystem dynamics. In: Veblen TT, Baker WL, Montenegro G, Swetnam TW (eds) Fire and global change in temperate ecosystems of the Western Americas, vol 160 (Ecological Studies). Springer-Verlag, New York, pp 32–68

Kreye JK, Kobziar LN, Camp JM (2014) Immediate and short-term response of understory fuels following mechanical mastication in a pine flatwoods site of Florida, USA. For Ecol Manage 313:340–354. doi:http://dx.doi.org/10.1016/j.foreco.2013.10.034

Liodakis S, Bakirtzis D, Dimitrakopoulos A (2002) Ignition characteristics of forest species in relation to thermal analysis data. Thermochim Acta 390(1–2):83–91. doi:http://dx.doi.org/10.1016/S0040-6031(02)00077-1

Mak EHT (1988) Notes: measuring foliar flammability with the limiting oxygen index method. For Sci 34(2):523–529

North M, Collins BM, Stephens S (2012) Using fire to increase the scale, benefits, and future maintenance of fuels treatments. J For 110 (7):392–401. doi:10.5849/jof.12-021

NWCG (2006) Fire terminology. http://www.nwcg.gov/pms/pubs/glossary/. Accessed 1 June 2014

Parsons RA, Mell WE, McCauley P (2010) Linking 3D spatial models of fuels and fire: effects of spatial heterogeneity on fire behavior. Ecol Model 222(3):679–691
Philpot CW (1969) Seasonal changes in heat content and ether extractive content of chamise. Res. Pap. U.S. Department of Agriculture, Forest Service, Intermountain Forest and Range Experiment Station, Research Paper INT-61, Ogden, UT, 10 pp
Pollet J, Omi PN (2002) Effect of thinning and prescribed burning on crown fire severity in ponderosa pine forests. Int J Wildland Fire 11:1–10
Reinhardt ED, Keane RE, Calkin DE, Cohen JD (2008) Objectives and considerations for wildland fuel treatment in forested ecosystems of the interior western United States. For Ecol Manage 256(12):1997–2006. doi:10.1016/j.foreco.2008.09.016
Schoennagel T, Veblen TT, Negron JF, Smith JM (2012) Effects of mountain pine beetle on fuels and expected fire behavior in lodgepole pine forests, Colorado, USA. PLoS ONE 7(1):e30002
Thaxton JM, Platt WJ (2006) Small-scale fuel variation alters fire intensity and shrub abundance in a pine savanna. Ecology 87(5):1331–1337. doi:10.1890/0012-9658(2006)87[1331:SFVAFI]2.0.CO;2

Glossary, Abbreviations and Acronyms[1]

Biomass. Dead and live organic material.

Burnable canopy biomass. Biomass consisting of all canopy fuel particles less than 3 mm in diameter.

Canopy bulk density (CBD). The amount of burnable canopy biomass in a unit volume of space.

Canopy base height (CBH). The height above the ground at which the canopy bulk density first exceeds a threshold value.

Canopy height (CH). The height above the ground at which the canopy bulk density last exceeds a threshold value.

Classification. A quantitative description of fuelbeds to simplify fuel attribute inputs for fire software applications.

Computational fluid dynamics (CFD). Term used to describe the next generation of fire models that simulate fire behavior using conservation of mass and momentum theory coupled with a computational fluid dynamics approach.

Coarse woody debris (CWD). Downed dead woody fuel particles that are greater than 8 cm in diameter. CWD does not include snags in this book. CWD is often synonymous with logs or 1000 h woody fuels.

Digital elevation model (DEM). A raster data layer of elevation above mean sea level assigned to each pixel in the raster.

Duff. Layer of decomposing organic materials and particles that are not readily identifiable lying just below the litter layer and immediately above the mineral soil

Endogenous. Originating inside the area of concern.

Exogenous. Originating outside of the area of concern.

Fine woody debris (FWD). The downed, dead woody fuel particles that are less than 8 cm in diameter.

Fire behavior fuel model (FBFM). A set of input fuel parameters for fire behavior models that have been adjusted to match observed fire behavior.

[1] This is a list of important terms, acronyms, and abbreviations used in this book. Definitions are given as used in this book even though some of the acronyms have multiple definitions in the literature.

R. E. Keane, *Wildland Fuel Fundamentals and Applications*,
DOI 10.1007/978-3-319-09015-3

Fire model. A set of algorithms that simulate an aspect of fire behavior or fire effects.

Fire modeling systems. Specially designed fire modeling software applications built to simulate fire behavior and effects using a combination of fire models.

First order fire effects model (FOFEM). A computer model to predict the immediate effects of a fire, specifically tree mortality, fuel consumption, and soil heating.

Fixed area plots (FAP). Using plot frames of fixed areas to sample fuel loadings.

Flammability. Ability of a fuel particle, type, component, or layer to ignite and combust

Fuel characteristics classification system (FCCS). A fuel classification system that offers consistently organized fuels data along with numerical inputs to fire behavior, fire effects, and dynamic vegetation models

Fuel loading models (FLMs). A classification of fuelbeds where the classes were identified using statistical clustering of estimates of smoke and soil heating simulated from the field-sampled loadings of the eight fuel components that comprise each fuelbed.

Fuel treatment longevity. The length of time (yrs) it takes for a fuel treatment to become ineffective.

Geographical information system (GIS). Computer software that allows for the spatial representation of various entities.

Herbs. All forbs, grasses, sedges, rushes, ferns and other non-woody vascular plant lifeforms. Mosses, lichens, and other non-vascular plants can be added depending on management objectives.

LANDFIRE. A national U.S. project to develop fine scale GIS layers for fire management

Leaf area index (LAI). The total surface area of all the leaves or needles for all the plants in plot divided by the area of that plot (m m^{-1})

LiDAR. A remote sensing technology that measures distance by hitting a target with a laser and analyzing various aspects of the reflected light. LiDAR is a portmanteau of "light" and "radar"

Loading. The mass of burnable organic biomass per unit area for any fuel component.

Log. a downed dead fuel particle that is larger than 8 cm (3 in) in diameter.

National Wildfire Coordination Group (NWCG). A group of fire professionals from various government agencies in the US that decide on the standards and terminology for fire research and management.

Particle density. The dry mass per unit volume of a fuel particle often expressed with the units kg m^{-3}.

Perpendicular distance sampling (PDS). A fuel sampling method for coarse woody debris that uses probablility proportional to volume concepts.

Photoload sampling (photoload). Visual estimation of fuel component loading using a set of pictures that show graduated fuel loadings where the conditions in observed in the field are matched with the most appropriate picture.

Planar intercept (PI). A sampling method where the counts or measurements of woody fuel particles that intersect a two dimensional sampling plane are used to compute woody fuel loading.

Potential vegetation type (PVT). A potential vegetation type is the vegetation type that would eventually dominate an area in the absence of disturbance.

Shrubs. Any woody, non-tree species. The selection of what constitutes shrubs is determined by management objective; vines, sub-shrubs, krumholtz, and other woody life forms can be included in shrubs if designated by the objective.

S.M.A.R.T. Objectives. A comprehensive way to specify fuel sampling objectives. SMART stands for Specific, Measureable, Achievable, Relevant, and Time-based

Snag. A standing dead tree that is sometimes considered coarse woody debris (CWD) and part of the woody fuel component.

Specific Gravity. The ratio of the density of a fuel particle relative to the density of water at a specific temperature, pressure.

Surface-Area-to-Volume Ratio (SAVR). The ratio of the area of the surface of a fuel particle(s) divided by the volume of that same fuel particle(s). SAVR has units of ft^{-1} or m^{-1}

Synthetic aperature radar (SAR). An active sensor

Terrestrial scanning lidar (TSL). A ground-based lidar sensor that scans an area and creates a three dimensional image of return distances for localized mapping of fuel.

Timelag. A term used to define the length of time needed to dry a woody fuel particle approximately 63 % of its moisture. For a 1 h timelag woody fuel, it would take 1 h to dry the fuel by 63 % of the starting moisture content.

Woody fuels. Any downed, dead organic biomass that originates from trees, shrubs, and other dendrites.

Index

A
Aboveground standing crop, 84
Abstraction, 119, 121
  drawbacks of, 120
Association, 112, 157
  approach, 133, 160
  biomass is fuel, 3
  disadvantage of, 161
  problems associated with, 114

B
Biomass, 3
  aboveground standing, 84
  production of, 83, 84
Biophysical modeling, 157, 165
Bulk density, 4, 6, 21, 31, 53, 142

C
Canopy
  base height, 59, 164
  bulk density, 31, 51, 60, 63, 66, 164
  burnable biomass, 22, 57
  closure, 59
  cover, 61, 68, 136, 137, 141, 142
  fuels, 4–6, 9, 31, 51, 57–59, 62, 65, 139, 140, 145
  opaqueness, 65
  profile, 31, 62–64, 66, 67, 164
Classification, 10, 44, 54, 117
  abstract, 120
  categories, 114
  of FBFM, 157
  of fuel, 111, 112, 115, 133, 134, 136, 154, 156
  vegetation, 161
  vegetation system, 155

D
Data
  ancillary, 155, 163
  field, 68, 111, 117, 126, 154, 157, 160, 162, 166–169
  spatial, 117, 155, 159, 160, 165, 168
Decomposition, 182
  concept of, 89
  Duff, 52
  gradient of, 41
  properties of, 90
  rate of, 12, 175
  various states of, 27
Deposition, 12, 18, 44, 182
  definition of, 87
  fuel, 93
  patterns of, 89
  rate of, 87
  studies on, 90
Disease, 7, 51, 63, 91, 96–98, 166
Disturbance, 7, 57, 83, 85–87, 91, 111, 114, 132, 159, 166, 175, 177, 179, 182
  adaptations, 18
  history, 27
  origin, 51
  regime, 47
Duff, 5, 28, 41, 52, 75, 120, 134, 136, 139, 140, 164, 178
  layers, 90
  properties, 53

F
Field assessment, 153, 159, 160
Fire
  behavior, 3, 4, 8–12, 15, 18, 19, 23–25, 29, 33, 41, 53, 59, 62, 67, 71, 114, 119, 120
  crown, 7, 31, 57, 59–61, 63, 66

R. E. Keane, *Wildland Fuel Fundamentals and Applications*,
DOI 10.1007/978-3-319-09015-3

active, 59–61, 63, 66
passive, 59–61
danger, 9, 11, 27, 120, 153
effects, 12, 23, 30, 61, 112, 118, 121, 178, 181
exclusion, 63, 180
ground, 9
hazard, 51, 59, 153, 178, 181
intensity, 8, 19, 28, 31, 33, 44, 48, 59, 63, 65, 79, 177, 180
mixed severity, 9
risk, 48, 126, 163
severity, 9, 168
spread, 4, 8, 9, 11, 17, 19, 22, 23, 33, 41, 44, 48, 50, 53, 60, 61, 63, 65, 79
stand-replacement, 9
suppression, 54, 80, 179
Flammability, 7, 30, 47, 117, 175, 176, 179
Fragmentation, 26, 72
Fuel
activity, 7
canopy, 4–6, 9, 31, 50, 57, 59, 61–63, 65, 67, 68, 145, 161, 163, 167, 177, 180
complex, 39, 122, 137
component, 6, 21, 22, 25, 27, 30, 31, 39, 41, 43, 44, 47, 50, 53, 71, 80, 81, 111, 112, 114, 115, 119, 125, 126, 128, 129, 133, 135–137, 140, 142, 143, 145, 146, 156, 164, 165, 168
dead, 6, 29, 30, 71–74, 77, 81
depth, 51, 112
flammability, 7, 176
fuel bed, 4–7, 11, 12, 21, 22, 27, 30, 33, 41, 51, 53, 111, 114, 115, 117–119, 121, 136, 141, 146, 154, 162, 182
ground, 5, 28, 39, 53, 77, 137
ladder, 7, 59, 66, 67
layer, 136, 137, 153, 168, 180
live, 6, 27, 43, 72, 76–78, 81, 147, 176
loading, 19, 22, 25, 31, 39, 66, 111, 112, 114, 119, 121, 127, 129, 133, 134, 136, 140, 142, 143, 147, 155, 157, 166, 168, 180, 181
models, 29, 33, 119, 120, 155, 161, 163
moisture content, 22, 71, 181
moisture of extinction, 22, 27, 79
particle, 6, 7, 12, 24–26, 28, 29, 32, 33, 43, 44, 47, 52, 62, 72–74, 80, 121, 136, 138, 142, 146
sampling, 25, 54, 65, 121, 125–128, 136, 137, 141, 145, 146, 160, 168, 182
surface, 4, 7, 9, 25, 32, 39, 41, 48, 50, 51, 53, 59, 67, 125, 129, 140, 143, 163, 178, 179, 181
treatment, 7, 12, 17, 53, 67, 114, 121, 126, 177–179
treatment longevity, 179
type, 5, 6, 9, 11, 27, 39, 44, 48
woody, 6, 7, 12, 25, 27, 39, 41, 43, 44, 75, 80, 139, 146, 165, 180
Fuel bed
depth, 21, 32, 34, 178
packing ratio, 33

G
Grazing, 34
Gross primary productivity, 84

H
Heat content, 6, 7, 19, 28, 29, 52, 117, 176
Herbaceous, 6, 39, 47, 50, 135
Herbs, 47, 136, 141, 164
Hydroscopy, 72

I
Imagery, 112, 133, 154, 155, 162–165, 167
technique, 137
Infiltration, 72, 73
Insects, 51, 57, 168
Inventory, 12, 41, 111, 118, 126, 133, 138, 141, 143–145

L
Landsat, 161–163, 167
Landscape
ecology, 182
metrics, 99
patch, 101
Lichens, 6, 48, 57
LiDAR, 162–164
Litter, 5, 7, 31, 40, 41, 75
Litter fall, 18, 40
Loading, 6, 10, 19, 22, 25, 30, 31, 41, 61, 118, 128, 134, 137, 141

M
Macro plots, 127, 140, 146
Mastication, 7, 51–53, 178
Micro plots, 127, 140, 146
Monitoring, 126, 129, 138
Mosses, 6, 46, 48, 57

N
Nan plots, 127, 146
Necromass, 6, 53, 147
Needle drape, 50
Net ecosystem production, 85
Net primary production, 84

O
Opportunistic
  approach, 115, 116
  classification, 116, 121
  flora, 96
Organic soils, 54, 75

P
Particle
  density, 22, 26, 27, 31, 44, 72, 140, 176
  diameter, 12, 21, 25, 43, 138, 146
Patch size, 99, 100
Photosynthesis, 10, 15, 28, 68, 76

R
Remote sensing, 154, 155, 157, 162, 165
  advantages of, 164
  ecological limitations of, 164
Respiration
  autotrophic, 84–86
  heterotrophic, 85
Roots, 54, 178
  role in water transport, 76

S
Sampling
  area, 136, 142
  associative, 133, 134, 161, 167
  cover and volume, 141
  cover-volume, 136, 142
  destructive, 62, 141–143, 145
  direct methods of, 138, 141
  fixed area plots, 127, 139, 142, 143
  gravimetric, 79, 80
  imagery, 162–164
  indirect methods, 133
  intensity, 125, 128
  perpendicular distance, 140
  photo, 115, 135
  photo series, 115, 116, 134, 135, 145
  photoloads, 135, 140, 143
  unit, 127, 129, 139, 146
  visual, 134, 136, 142, 143, 145
Satellite
  imagery, 112, 137, 155, 160, 163, 167
Scat, 50
Semivariograms, 100
Sensors
  active, 162
  passive, 162
Shade tolerant
  species, 86, 177
Shrubs, 6, 33, 46, 50, 77, 136, 141
Slash, 7, 28, 139, 180
Snag, 3, 48, 52
Squirrel middens, 50
Stomates, 76, 78
Stumps, 3, 52
Succession, 17
Surface area to volume ratio (SAVR), 7, 47

T
Terrestrial scanning lidar (TSL), 137, 144
Thermatic mapper, 167
Timelag
  concept of, 43, 74
Transpiration, 6, 68, 76
Tree
  crown, 7, 57, 78
  crown height, 62
  height, 57, 66, 144
  live crown ratio, 144
  regeneration, 12, 31, 50, 179
  seedlings, 7, 50

V
Variability
  spatial, 65, 134, 182
  temporal, 121
Vegetation, 6, 7, 11, 46, 51
  ancillary, 116
  based measurement, 100
  characteristics, 114
  development, 17, 27, 180
  dynamics, 4
  succession, 86

W
Wetlands, 54
Wind, 7, 21, 33, 51
  dispersal using, 89
  light, 87
  properties of, 93

MIX
Papier aus verantwortungsvollen Quellen
Paper from responsible sources
FSC® C105338

If you have any concerns about our products,
you can contact us on
ProductSafety@springernature.com

In case Publisher is established outside the EU,
the EU authorized representative is:
Springer Nature Customer Service Center GmbH
Europaplatz 3, 69115 Heidelberg, Germany

Printed by Libri Plureos GmbH
in Hamburg, Germany